THE EMERGENCE OF NUMERICAL WEATHER PREDICTION
Richardson's Dream

In the early twentieth century, Lewis Fry Richardson dreamt that scientific weather prediction would one day become a practical reality. The method of computing changes in the state of the atmosphere that he mapped out in great detail is essentially the method used today. Before his ideas could bear fruit several advances were needed: better understanding of the dynamics of the atmosphere; stable computational algorithms to integrate the equations of motion; regular observations of the free atmosphere; and powerful automatic computer equipment.

By 1950, advances on all these fronts were sufficient to permit the first computer weather forecast to be made. Over the ensuing 50 years progress in numerical weather prediction has been dramatic, allowing Richardson's dream to become a reality. Weather prediction and climate modelling have now reached a high level of sophistication.

This book tells the story of Richardson's trial forecast, and the fulfilment of his dream of practical weather forecasting and climate modelling. It has a complete reconstruction of Richardson's forecast, and analyses in detail the causes of the failure of this forecast. It also includes a description of current practice, with particular emphasis on the work of the European Centre for Medium-Range Weather Forecasts. This book will appeal to everyone involved in numerical weather forecasting, from researchers and graduate students to professionals.

PETER LYNCH is Met Éireann Professor of Meteorology at University College Dublin (UCD) and Director of the UCD Meteorology and Climate Centre. Prior to this he was Deputy Director of Met Éireann, the Irish Meteorological Service. He is a Fellow of the Royal Meteorological Society, the Royal Astronomical Society, the Institute of Mathematics and its Applications, and the Institute of Physics.

THE EMERGENCE OF NUMERICAL WEATHER PREDICTION

Richardson's Dream

PETER LYNCH
University College Dublin

CAMBRIDGE
UNIVERSITY PRESS

32 Avenue of the Americas, New York NY 10013-2473, USA

Cambridge University Press is part of the University of Cambridge.

It furthers the University's mission by disseminating knowledge in the pursuit of
education, learning and research at the highest international levels of excellence.

www.cambridge.org
Information on this title: www.cambridge.org/9781107414839

First published 2006
First paperback edition 2014

A catalogue record for this publication is available from the British Library

ISBN 978-0-521-85729-1 Hardback
ISBN 978-1-107-41483-9 Paperback

Contents

Guiding signs

Richardson was considerate in providing 'guiding signs' to assist readers in navigating his book, *Weather Prediction by Numerical Process*. We follow his example.

  (i) *Weather Prediction by Numerical Process* is denoted *WPNP* throughout.
 (ii) In general, quantities are given in SI units. However, Richardson used the older CGS system and, where apppropriate, values of quantities are given in this system, with due indication.
(iii) Equations are numbered sequentially within each Chapter. Thus, the reference (2.20) denotes equation number 20 in Chapter 2.
 (iv) The generic term 'gravity waves' is used for pure gravity waves, gravity-inertia waves and acoustic-inertia waves.
  (v) Historians of meteorology may find Chapters 1, 3, 6, 7 and 10 of primary interest.
 (vi) Chapters 2 and 3 are the most mathematically involved. On first reading they may be skimmed over, as the bulk of the remaining material should be accessible without detailed knowledge of them.
(vii) The current state of the science of numerical weather prediction is presented in Chapter 11.
(viii) A list of the main symbols is given in Appendix 1 (p. 251).

Preface

Accurate weather forecasts based on computer simulation of the atmosphere are now available routinely throughout the world. Numerical Weather Prediction (NWP) has developed rapidly over the past fifty years and the power of computer models to forecast the weather has grown impressively with the power of computers themselves. Earth System Models are capable of simulating climates of past millennia and are our best means of predicting future climate change, the major environmental threat facing humankind today.

It is remarkable that the basic techniques of numerical forecasting and climate modelling in use today were developed long before the first electronic computer was constructed. Lewis Fry Richardson first considered the problem of weather forecasting in 1911. He had developed a versatile technique for calculating approximate solutions of complicated mathematical equations – nonlinear partial differential equations – and he realised that it could be applied to the equations that govern the evolution of atmospheric flows. Recognising that a practical implementation of his method would involve a phenomenal amount of numerical calculation, Richardson imagined a fantastic forecast factory with a staff of thousands of human computers busily calculating the terms in the fundamental equations and combining their results in an ingeniously organised way to produce a weather forecast.

Richardson did much more than set down the principles of scientific weather prediction: he constructed, in complete detail, a systematic procedure or algorithm for generating the numerical solution of the governing equations. And he went further still: he applied the procedure to a real-life case and calculated the initial changes in pressure and wind. Although the resulting 'forecast' was unrealistic, Richardson's numerical experiment demonstrated that his procedure was self-consistent and, in principle, feasible. However, there were several major practical obstacles to be overcome before numerical prediction could be put into practice.

In this book we discuss Richardson's method in detail, showing how his numerical procedure is constructed by application of his finite differencing technique to the atmospheric equations. The resulting algorithm is ideally suited for programming on a modern computer. We describe the implementation of Richardson's algorithm as a computer program, and show that his forecast results can be replicated accurately. Richardson was meticulous and methodical; the consistency between the original and reconstructed predictions is more a demonstration of the validity of the computer program than a confirmation of the correctness of Richardson's calculations.

Although mathematically correct, Richardson's prediction was physically unrealistic. The reasons for this are considered in depth. The core of the problem is that a delicate dynamic balance that prevails in the atmosphere was not reflected in the initial data used by Richardson. The consequence of the imbalance was the contamination of the forecast by spurious noise. Balance may be restored by small but critical adjustments to the data. We show that after such modification, called initialisation, a physically realistic forecast is obtained. This further demonstrates the integrity of Richardson's process. We examine his discussion on smoothing the initial data, relating it to the initialisation procedure and showing how close he came to overcoming the difficulties in his forecast.

The true significance of Richardson's work was not immediately evident; the computational complexity of the process and the disastrous results of the single trial forecast tended to deter others from following the trail mapped out by him. But his work was of key importance to the pioneers who carried out the first automatic forecast on an electronic computer, in 1950. One of them, Jule Charney, addressing the Royal Meteorological Society some years later, said 'to the extent that my work in weather prediction has been of value, it has been a vindication of the vision of my distinguished predecessor, Lewis F. Richardson'.

Richardson expressed a dream that, 'some day in the dim future', numerical weather prediction would become a practical reality. Progress was required on several fronts before this dream could be realised. A fuller understanding of atmospheric dynamics allowed the development of simplified systems of equations; regular radiosonde observations of the free atmosphere and, later, satellite data, provided the initial conditions; stable finite difference schemes were developed; and powerful electronic computers provided a practical means of carrying out the prodigious calculations required to predict the changes in the weather.

Progress in weather forecasting and in climate modelling over the past 50 years has been dramatic. We review the current status of numerical prediction, both deterministic and probabilistic, with particular emphasis on the work of the European Centre for Medium-Range Weather Forecasts. Richardson's remarkable prescience and the abiding value of his work are illustrated by the wide range of ideas, central to modern weather and climate models, that originated with him. Thus, it may be reasonably claimed that his work is the basis of modern weather and climate forecasting.

Acknowledgements

It has taken many years to complete this project and, in the process, it has been my privilege to work with a wide range of great people. To those not mentioned below, and to 'the unknown meteorologist', I offer my profound gratitude.

My interest in Richardson's forecast stems from discussions with my doctoral adviser and colleague, Ray Bates. Ray has provided advice and assistance on a multitude of meteorological matters. The idea of initialising Richardson's data and repeating his forecast occurred to me when I first read George Platzman's learned and comprehensive overview of *Weather Prediction by Numerical Process*. Written after the appearance of the Dover edition, this article has been a continuing source of inspiration. I greatly enjoyed visiting Professor Platzman in Chicago and have derived much assistance from correspondence with him, particularly relating to the introductory chapter and the section on the ENIAC integrations.

I first read Oliver Ashford's biography of Richardson while a visiting scientist at the Royal Netherlands Meteorological Institute (KNMI) in 1985, and I recall several fruitful and fascinating conversations about it with my Dutch colleagues. Following the appearance of my article on Richardson's barotropic forecast, in 1992, Oliver invited me to contact him. I have since enjoyed several visits to his home and have benefited from many conversations with him about Richardson and his work. Thanks also to Oliver for providing a photograph of Richardson.

Serious work on the reconstruction of the forecast began in Spring 1991 during a visit to the International Meteorological Institute (IMI) in Stockholm. I am grateful to Erland Källén for inviting me to visit IMI. While there, I analysed the initial mass fields for the forecast while Élias Hólm analysed the winds. Work on digital filtering initialisation, which I began with Xiang-Yu Huang in Stockholm, was also crucial for the success of the project.

A number of colleagues were kind enough to read and comment on chapters of the book. I am especially grateful to Akira Kasahara for his extensive review of an early draft. I also enjoyed and benefited greatly from discussions with Akira during a visit to the National Center for Atmospheric Research (NCAR); the visit was facilitated by Joe Tribbia. Sections of the chapter on operations at the European Centre were reviewed by Tim Palmer, Adrian Simmons and Sakari Uppala, and I thank them for their comments.

Support from the Director of Met Éireann, Declan Murphy, and from my many colleagues at Met Éireann, is acknowledged with warm gratitude. Jim Hamilton provided willing and substantial assistance in the preparation of graphics. Aidan McDonald provided a copy of his spherical shallow water code. Jim Logue reviewed early drafts of some chapters. Lisa Shields, Librarian at Met Éireann, and her successor Jane Burns, assisted with translations and literature searches. Eoin Moran and Jackie O Sullivan provided welcome hospitality during my stay at Valentia Observatory. Warmest thanks to them and to all my other colleagues in Met Éireann.

An office at the School of Cosmic Physics in Dublin was provided during the Summer of 2003 by the Dublin Institute for Advanced Studies, allowing me to work undisturbed by quotidian chores. I thank Peter Readman for arranging this accommodation and all the staff at 'Number 5' for their hospitality.

I enjoyed visiting a number of libraries and archives, and was received with un-failing courtesy and enthusiasm. I am particularly grateful to: Graham Bartlett (Met Office Archives), Elisabeth Kaplan (Charles Babbage Institute, Minnesota), Nora Murphy (MIT Archives), Jenny O'Neill (MIT Museum), Diane Rabson (NCAR Archives) and Godfrey Waller (Cambridge University Library).

I have been fortunate to have received assistance from many people in ways that they may have forgotten but that I remember with gratitude. Their help ranged from mild interest in the project to the provision of substantial assistance. With apologies to others whom I may have omitted, I thank: Michael Börngen, Huw Davies, Terry Davies, Dezso Devenyi, Jean-François Geleyn, Sigbjørn Grønås, Tony Hollingsworth, Brian Hoskins, Jean-Pierre Javelle, Eugenia Kalnay, Leif Laursen, Andrew Lorenc, Ray McGrath, Michael McIntyre, Detlev Majewski, Brendan McWilliams, Fedor Mesinger, Anders Persson, Norman Phillips, Ian Roulstone, Richard Swinbank, Per Undén, Hans Volkert, Bruce Webster, Andy White and Dave Williamson.

My students Michael Clark and Paul Nolan provided assistance in proofreading and preparation of figures. I am grateful to Joy Davies, Haldo Vedin and Christer Kiselman for assistance in translation for the Table of notation in Appendix 1. The staff of Cambridge University Press, especially Matt Lloyd and Emma Pearce, have been very helpful.

I am grateful to my sons, Owen and Andrew, for help in overcoming a number of perverse problems with computers and for scanning images, and to my wife, Cabrini, for her unwavering love and support.

1

Weather Prediction by Numerical Process

> Perhaps some day in the dim future it will be possible to advance the
> computations faster than the weather advances and at a cost less than the
> saving to mankind due to the information gained. But that is a dream.
>
> *(WPNP, p. vii; Dover Edn., p. xi)*

Lewis Fry Richardson's extraordinary book *Weather Prediction by Numerical Process*, published in 1922, is a strikingly original scientific work, one of the most remarkable books on meteorology ever written. In this book – which we will refer to as *WPNP* – Richardson constructed a systematic mathematical method for predicting the weather and demonstrated its application by carrying out a trial forecast. History has shown that his innovative ideas were fundamentally sound: the methodology proposed by him is essentially that used in practical weather forecasting today. However, the method devised by Richardson was utterly impractical at the time of its publication, and the results of his trial forecast appeared to be little short of outlandish. As a result, his ideas were eclipsed for decades and his wonderful opus gathered dust and was all but forgotten.

1.1 The problem

Imagine you are standing by the ocean shore, watching the sea rise and fall as wave upon wave breaks on the rocks. At a given moment the water is rising at a rate of one metre per second – soon it will fall again. Is there an ebb or a flood tide? Suppose you use the observed rate of change and extrapolate it over the six hours that elapse between tidal extremes; you will obtain an extraordinary prediction: the water level should rise by some 20 km, twice the height of Mt Everest. This forecast is meaningless! The water level is governed by physical processes with a wide range of timescales. The tidal variations, driven by lunar gravity, have a period of around 12 hours, linked to the Earth's rotation. But wind-driven waves and swell

1

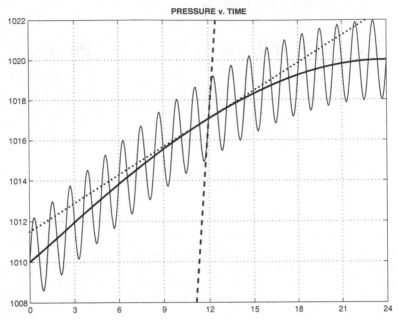

Figure 1.1 Schematic illustration of pressure variation over a 24 hour period. The thick line is the mean, long-term variation, the thin line is the actual pressure, with high frequency noise. The dotted line shows the rate of change, at 12 hours, of the mean pressure, and the dashed line shows the corresponding rate of change of the actual pressure. (After Phillips, 1973)

vary on a timescale of seconds. The instantaneous change in level due to a wave is no guide to the long-term tidal variations: if the observed rise is extrapolated over a period much longer than the timescale of the wave, the resulting forecast will be calamitous.

In 1922, Richardson presented such a forecast to the world. He calculated a change of atmospheric pressure, for a particular place and time, of 145 hPa in 6 hours. This was a totally unrealistic value, too large by two orders of magnitude. The prediction failed for reasons similar to those that destroy the hypothetical tidal forecast. The spectrum of motions in the atmosphere is analogous to that of the ocean: there are long-period variations dominated by the effects of the Earth's rotation – these are the meteorologically significant rotational modes – and short-period oscillations called gravity waves, having speeds comparable to that of sound. The interaction between the two types of variation is weak, just as is the interaction between wind-waves and tidal motions in the ocean; and, for many purposes, the gravity waves, which are normally of small amplitude, may be treated as irrelevant noise.

Although they have little effect on the long-term evolution of the flow, gravity waves may profoundly influence the way it changes on shorter timescales. Figure 1.1

(after Phillips, 1973) schematically depicts the pressure variation over a period of one day. The smooth curve represents the variation due to meteorological effects; its gentle slope (dotted line) indicates the long-term change. The rapidly varying curve represents the actual pressure changes when gravity waves are superimposed on the meteorological flow: the slope of the oscillating curve (dashed line) is precipitous and, if used to determine long-range variations, yields totally misleading results. What Richardson calculated was the instantaneous rate of change in pressure for an atmospheric state having gravity-wave components of large amplitude. This tendency, $\partial p / \partial t \approx 0.7\,\mathrm{Pa\,s}^{-1}$, was a sizeable but not impossible value. Such variations are observed over short periods in intense, localised weather systems.[1] The problem arose when Richardson used the computed value in an attempt to deduce the long-term change. Multiplying the calculated tendency by a time step of six hours, he obtained the unacceptable value quoted above. The cause of the failure is this: *the instantaneous pressure tendency does not reflect the long-term change.*

This situation looks hopeless: how are we to make a forecast if the tendencies calculated using the basic equations of motion do not guide us? There are several possible ways out of the dilemma; their success depends crucially on the decoupling between the gravity waves and the motions of meteorological significance – we can distort the former without seriously corrupting the latter.

The most obvious approach is to construct a forecast by combining many time steps which are short enough to enable accurate simulation of the detailed high-frequency variations depicted schematically in Fig. 1.1. The existence of these high-frequency solutions leads to a stringent limitation on the size of the time step for accurate results; this limitation or *stability criterion* was discovered in a different context by Hans Lewy in Göttingen in the 1920s (see Reid, 1976), and was first published in Courant *et al.* (1928). Thus, although these oscillations are not of meteorological interest, their presence severely limits the range of applicability of the tendency calculated at the initial time. Small time steps are required to represent the rapid variations and ensure accuracy of the long-term solution. If such small steps are taken, the solution will contain gravity-wave oscillations about an essentially correct meteorological flow. One implication of this is that, if Richardson could have extended his calculations, taking a large number of small steps, his results would have been noisy but the mean values would have been meteorologically reasonable (Phillips, 1973). Of course, the attendant computational burden made this impossible for Richardson.

The second approach is to modify the governing equations in such a way that the gravity waves no longer occur as solutions. This process is known as filtering

---

[1] For example, Loehrer and Johnson (1995) reported a surface pressure drop of 4 hPa in five minutes in a mesoscale convective system, or $\partial p / \partial t \approx -1.3\,\mathrm{Pa\,s}^{-1}$.

the equations. The approach is of great historical importance. The first successful computer forecasts (Charney *et al.*, 1950) were made with the barotropic vorticity equation (see Chapter 10), which has low-frequency but no high-frequency solutions. Later, the quasi-geostrophic equations were used to construct more realistic filtered models and were used operationally for many years. An interesting account of the development of this system appeared in Phillips (1990). The quasi-geostrophic equations are still of great theoretical interest (Holton, 2004) but are no longer considered to be sufficiently accurate for numerical prediction.

The third approach is to adjust the initial data so as to reduce or eliminate the gravity-wave components. The adjustments can be small in amplitude but large in effect. This process is called *initialisation*, and it may be regarded as a form of smoothing. Richardson realised the requirement for smoothing the initial data and devoted a chapter of *WPNP* to this topic. We will examine several methods of initialisation in this work, in particular in Chapter 8, and will show that the digital-filtering initialisation method yields realistic tendencies when applied to Richardson's data.

The absence of gravity waves from the initial data results in reasonable initial rates of change, but it does not automatically allow the use of large time steps. The existence of high-frequency solutions of the governing equations imposes a severe restriction on the size of the time step allowable if reasonable results are to be obtained. The restriction can be circumvented by treating those terms of the equations that govern gravity waves in a numerically implicit manner; this distorts the structure of the gravity waves but not of the low-frequency modes. In effect, implicit schemes slow down the faster waves thus removing the cause of numerical instability (see §5.2 below). Most modern forecasting models avoid the pitfall that trapped Richardson by means of initialisation followed by semi-implicit integration.

1.2 Vilhelm Bjerknes and scientific forecasting

At the time of the First World War, weather forecasting was very imprecise and unreliable. Observations were scarce and irregular, especially for the upper air and over the oceans. The principles of theoretical physics played a relatively minor role in practical forecasting: the forecaster used crude techniques of extrapolation, knowledge of climatology and guesswork based on intuition; forecasting was more an art than a science. The observations of pressure and other variables were plotted in symbolic form on a weather map and lines were drawn through points with equal pressure to reveal the pattern of weather systems – depressions, anticyclones, troughs and ridges. The concept of *fronts,* surfaces of discontinuity between warm and cold airmasses, had yet to emerge. The forecaster used his experience, memory

Figure 1.2 A recent painting (from photographs) of the Norwegian scientist Vilhelm Bjerknes (1862–1951) standing on the quay in Bergen. (© *Geophysical Institute, Bergen*. Artist: Rolf Groven)

of similar patterns in the past and a menagerie of empirical rules to produce a forecast map. Particular attention was paid to the reported pressure changes or tendencies; to a great extent it was assumed that what had been happening up to now would continue for some time. The primary physical process attended to by the forecaster was *advection*, the transport of fluid characteristics and properties by the movement of the fluid itself.

The first explicit analysis of the weather-prediction problem from a scientific viewpoint was undertaken at the beginning of the twentieth century when the Norwegian scientist Vilhelm Bjerknes set down a two-step plan for rational forecasting (Bjerknes, 1904):

If it is true, as every scientist believes, that subsequent atmospheric states develop from the preceding ones according to physical law, then it is apparent that the necessary and sufficient conditions for the rational solution of forecasting problems are the following:
1. A sufficiently accurate knowledge of the state of the atmosphere at the initial time.
2. A sufficiently accurate knowledge of the laws according to which one state of the atmosphere develops from another.

Bjerknes used the medical terms *diagnostic* and *prognostic* for these two steps (Friedman, 1989). The diagnostic step requires adequate observational data to define the three-dimensional structure of the atmosphere at a particular time. There was a severe shortage of observations, particularly over the seas and for the upper air, but Bjerknes was optimistic:

We can hope ... that the time will soon come when either as a daily routine, or for certain designated days, a complete diagnosis of the state of the atmosphere will be available. The first condition for putting forecasting on a rational basis will then be satisfied.

In fact, such designated days, on which upper air observations were made throughout Europe, were organised around that time by the International Commission for Scientific Aeronautics.

The second, or prognostic, step was to be taken by assembling a set of equations, one for each dependent variable describing the atmosphere. Bjerknes listed seven basic variables: pressure, temperature, density, humidity and three components of velocity. He then identified seven independent equations: the three hydrodynamic equations of motion, the continuity equation, the equation of state and the equations expressing the two laws of thermodynamics. (As pointed out by Eliassen (1999), Bjerknes was in error in listing the second law of thermodynamics; he should instead have specified a continuity equation for water substance.) Bjerknes knew that an exact analytical integration was beyond our ability. His idea was to represent the initial state of the atmosphere by a number of charts giving the distribution of the variables at different levels. Graphical or mixed graphical and numerical methods, based on the fundamental equations, could then be applied to construct a new set of charts describing the state of the atmosphere, say, three hours later. This process could be repeated until the desired forecast length was reached. Bjerknes realised that the prognostic procedure could be conveniently separated into two stages, a purely hydrodynamic part and a purely thermodynamic part; the hydrodynamics would determine the movement of an airmass over the time interval and thermodynamic considerations could then be used to deduce changes in its state. He concluded:

It may be possible some day, perhaps, to utilise a method of this kind as the basis for a daily practical weather service. But however that may be, the fundamental scientific study of atmospheric processes sooner or later has to follow a method based upon the laws of mechanics and physics.

Bjerknes' speculations are reminiscent of Richardson's 'dream' of practical scientific weather forecasting.[2]

A tentative first attempt at mathematically forecasting synoptic changes by the application of physical principles was made by Felix Exner, working in Vienna. His account (Exner, 1908) appeared only four years after Bjerknes' seminal paper. Exner

---

[2] Bjerknes' ideas on rational forecasting were adumbrated by Cleveland Abbe. See note added in proof, p. 27.

makes no reference to Bjerknes' work, which was also published in *Meteorologische Zeitschrift*. Though he may be presumed to have known about Bjerknes' ideas, Exner followed a radically different line: whereas Bjerknes proposed that the full system of hydrodynamic and thermodynamic equations be used, Exner's method was based on a system reduced to the essentials. He assumed that the atmospheric flow is geostrophically balanced and that the thermal forcing is constant in time. Using observed temperature values, he deduced a mean zonal wind. He then derived a prediction equation representing advection of the pressure pattern with constant westerly speed, modified by the effects of diabatic heating. It yielded a realistic forecast in the case illustrated in Exner's paper. Figure 1.3 shows his calculated pressure change (top) and the observed change (bottom) over the four-hour period between 8 p.m. and midnight on 3 January 1895; there is reasonable agreement between the predicted and observed changes. However, the method could hardly be expected to be of general utility. Exner took pains to stress the limitations of his method, making no extravagant claims for it. But despite the very restricted applicability of the technique devised by him, the work is deserving of attention as a first attempt at systematic, scientific weather forecasting. Exner's numerical method was summarised in his textbook (Exner, 1917, §70). The only reference by Richardson to the method was a single sentence (*WPNP*, p. 43) 'F. M. Exner has published a prognostic method based on the source of air supply.' It would appear from this that Richardson was not particularly impressed by it!

In 1912, Bjerknes became the first Director of the new Geophysical Institute in Leipzig. In his inaugural lecture he returned to the theme of scientific forecasting. He observed that 'physics ranks among the so-called exact sciences, while one may be tempted to cite meteorology as an example of a radically inexact science'. He contrasted the methods of meteorology with those of astronomy, for which predictions of great accuracy are possible, and described the programme of work upon which he had already embarked: *to make meteorology into an exact physics of the atmosphere*. Considerable advances had been made in observational meteorology during the previous decade, so that now the diagnostic component of his two-step programme had become feasible.

... now that complete observations from an extensive portion of the free air are being published in a regular series, a mighty problem looms before us and we can no longer disregard it. We must apply the equations of theoretical physics not to ideal cases only, but to the actual existing atmospheric conditions as they are revealed by modern observations. These equations contain the laws according to which subsequent atmospheric conditions develop from those that precede them. It is for us to discover a method of practically utilising the knowledge contained in the equations. From the conditions revealed by the observations we must learn to compute those that will follow. The problem of accurate pre-calculation that was solved for astronomy centuries ago must now be attacked in all earnest for meteorology. *(Bjerknes, 1914a)*

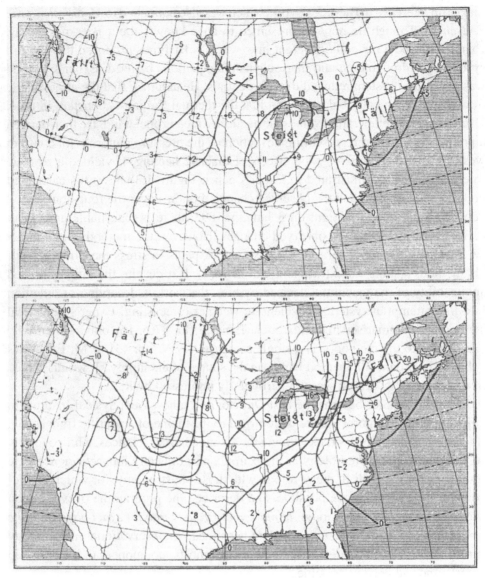

Figure 1.3 Top: Exner's calculated pressure change between 8 p.m. and midnight, 3 January 1895. Bottom: observed pressure change for the same period [Units: hundredths of an inch of mercury. *Steigt* = rises; *Fällt* = falls]. (Exner, 1908)

Bjerknes expressed his conviction that the acid test of a science is its utility in forecasting: 'There is after all but one problem worth attacking, *viz.*, the precalculation of future conditions.' He recognised the complexity of the problem and realised that a rational forecasting procedure might require more time than the atmosphere itself takes to evolve, but concluded:

If only the calculation shall agree with the facts, the scientific victory will be won. Meteorology would then have become an exact science, a true physics of the atmosphere. When that point is reached, *then* the practical results will soon develop.

It may require many years to bore a tunnel through a mountain. Many a labourer may not live to see the cut finished. Nevertheless this will not prevent later comers from riding through the tunnel at express-train speed.

At Leipzig, Bjerknes instigated the publication of a series of weather charts based on the data that were collected during the internationally-agreed intensive observation days and compiled and published by Hugo Hergesell in Strasbourg (these charts are discussed in detail in Chapter 6). One such publication (Bjerknes, 1914b), together with the 'raw data' in Hergesell (1913), was to provide Richardson with the initial conditions for his forecast.

Richardson first heard of Bjerknes' plan for rational forecasting in 1913, when he took up employment with the Meteorological Office. In the preface to *WPNP* he writes:

The extensive researches of V. Bjerknes and his School are pervaded by the idea of using the differential equations for all that they are worth. I read his volumes on *Statics* and *Kinematics* soon after beginning the present study, and they have exercised a considerable influence throughout it.

Richardson's book opens with a discussion of then-current practice in the Met Office. He describes the use of an Index of Weather Maps, constructed by classifying old synoptic charts into categories. The Index (Gold, 1920) assisted the forecaster to find previous maps resembling the current one and therewith to deduce the likely development by studying the evolution of these earlier cases:

The forecast is based on the supposition that what the atmosphere did then, it will do again now. There is no troublesome calculation, with its possibilities of theoretical or arithmetical error. The past history of the atmosphere is used, so to speak, as a full-scale working model of its present self. *(WPNP, p. vii; Dover Edn., p. xi)*

Bjerknes had contrasted the precision of astronomical prediction with the 'radically inexact' methods of weather forecasting. Richardson returned to this theme in his preface:

– the *Nautical Almanac,* that marvel of accurate forecasting, is not based on the principle that astronomical history repeats itself in the aggregate. It would be safe to say that a particular disposition of stars, planets and satellites never occurs twice. Why then should we expect a present weather map to be exactly represented in a catalogue of past weather? ... This alone is sufficient reason for presenting, in this book, a scheme of weather prediction which resembles the process by which the *Nautical Almanac* is produced, in so far as it is founded upon the differential equations and not upon the partial recurrence of phenomena in their ensemble.

Richardson's forecasting scheme amounts to a precise and detailed implementation of the prognostic component of Bjerknes' programme. It is a highly intricate procedure: as Richardson observed, 'the scheme is complicated because the atmosphere is complicated'. It also involved an enormous volume of numerical computation and was quite impractical in the pre-computer era. But Richardson was undaunted, expressing his dream that 'some day in the dim future it will be possible to advance the computations faster than the weather advances'. Today, forecasts are prepared routinely on powerful computers running algorithms that are remarkably similar to Richardson's scheme – his dream has indeed come true.

Before discussing Richardson's forecast in more detail, we will digress briefly to consider his life and work from a more general viewpoint.

1.3 Outline of Richardson's life and work

Richardson's life and work are discussed in a comprehensive and readable biography (Ashford, 1985). The Royal Society Memoir of Gold (1954) provides a more succinct description and the *Collected Papers* of Richardson, edited by Drazin (LFR I) and Sutherland (LFR II), include a biographical essay by Hunt (1993); see also Hunt (1998). Brief introductions to Richardson's work in meteorology (by Henry Charnock), in numerical analysis (by Leslie Fox) and on fractals (by Philip Drazin) are also included in Volume 1 of the *Collected Papers*. The article by Chapman (1965) is worthy of attention and some fascinating historical background material may be found in the review by Platzman (1967). In a recent popular book on mathematics, Körner (1996) devotes two chapters (69 pages) to various aspects of Richardson's mathematical work. The National Cataloguing Unit for the Archives of Contemporary Scientists has produced a comprehensive catalogue of the papers and correspondence of Richardson, which were deposited by Oliver Ashford in Cambridge University Library (NCUACS, 1993). The following sketch of Richardson's life is based primarily on Ashford's book.

Lewis Fry Richardson was born in 1881, the youngest of seven children of David Richardson and Catherine Fry, both of whose families had been members of the Society of Friends for generations. He was educated at Bootham, the Quaker school in York, where he showed an early aptitude for mathematics, and at Durham College of Science in Newcastle. He entered King's College, Cambridge in 1900 and graduated with a First Class Honours in the Natural Science Tripos in 1903. In 1909, he married Dorothy Garnett. They had no offspring but adopted two sons and a daughter, Olaf (1916–83), Stephen (1920–) and Elaine (1927–).

Over the ten years following his graduation, Richardson held several short research posts (Appendix 2 contains a chronology of the milestones of his life and career). As a scientist with National Peat Industries, he investigated the optimum

L. F. Richardson, 1931

Figure 1.4 Lewis Fry Richardson (1881–1953). Photograph by Walter Stoneman, 1931, when Richardson was aged 50. (Copy of photograph courtesy of Oliver Ashford)

method of cutting drains to remove water from peat bogs. The problem was formulated in terms of Laplace's equation on an irregularly-shaped domain. As this partial differential equation is not soluble by analytical means, except in special cases, he devised an approximate graphical method of solving it. More significantly, he then constructed a finite difference method for solving such systems and described this more powerful and flexible method in a comprehensive report (Richardson, 1910).

Around 1911, Richardson began to think about the application of his finite difference approach to the problem of forecasting the weather. He stated in the preface of *WPNP* that the idea first came to him in the form of a fanciful idea about a forecast factory, to which we will return in the final chapter. Richardson began serious work on weather prediction in 1913, when he joined the Met Office and was appointed Superintendent of Eskdalemuir Observatory, at an isolated location in Dumfrieshire in the Southern Uplands of Scotland. In May 1916, he resigned from the Met Office in order to work with the Friends Ambulance Unit (FAU) in France. There he spent over two years as an ambulance driver, working in close proximity to the fighting and on occasions coming under heavy shell fire. He returned to England after the cessation of hostilities and was employed once again by the Met Office to work at Benson, between Reading and Oxford, with W. H. Dines. The conditions of his employment included *experiments with a view to forecasting by numerical process.* He also developed several ingenious instruments for making upper air observations. However, he was there only one year when the Office came under the authority of the Air Ministry, which also had responsibility for the Royal Air Force and, as a committed pacifist, he felt obliged to resign once more.

Richardson then obtained a post as a lecturer in mathematics and physics at Westminister Training College in London. His meteorological research now focused primarily on atmospheric turbulence. Several of his publications during this period are still cited by scientists. In one of the most important – *The supply of energy from and to atmospheric eddies* (Richardson, 1920) – he derived a criterion for the onset of turbulence, introducing what is now known as the Richardson Number. In another (Richardson, 1926), he investigated the separation of initially proximate tracers in a turbulent flow, and arrived empirically at his 'four-thirds law': the rate of diffusion is proportional to the separation raised to the power $4/3$. This was later established more rigourously by Kolmogorov (1941) using dimensional analysis. Bachelor (1950) showed the consistency between Richardson's four-thirds law and Kolmogorov's similarity theory. A simple derivation of the four-thirds law using dimensional analysis is given by Körner (1996).

In 1926, Richardson was elected a Fellow of the Royal Society. Around that time he made a deliberate break with meteorological research. He was distressed that his turbulence research was being exploited for military purposes. Moreover, he had taken a degree in psychology and wanted to apply his mathematical knowledge in that field. Among his interests was the quantitative measurement of human sensation such as the perception of colour. He established for the first time a logarithmic relationship between the perceived loudness and the physical intensity of a stimulus. In 1929, he was appointed Principal of Paisley Technical College, near Glasgow, and he worked there until his retirement in 1940.

From about 1935 until his death in 1953, Richardson thrust himself energetically into *peace studies*, developing mathematical theories of human conflict and the causes of war. Once again he produced ideas and results of startling originality. He pioneered the application of quantitative methods in this extraordinarily difficult area. As with his work in numerical weather prediction, the value of his efforts was not immediately appreciated. He produced two books, *Arms and Insecurity* (1947), a mathematical theory of arms races, and *Statistics of Deadly Quarrels* (1950) in which he amassed data on all wars and conflicts between 1820 and 1949 in a systematic collection. His aim was to identify and understand the causes of war, with the ultimate humanitarian goal of preventing unnecessary waste of life. However, he was unsuccessful in finding a publisher for these books (the dates refer to the original microfilm editions). The books were eventually published posthumously in 1960, thanks to the efforts of Richardson's son Stephen. These studies continue to be a rich source of ideas. A recent review of Richardson's theories of war and peace has been written by Hess (1995).

Richardson's genius was to apply quantitative methods to problems that had traditionally been regarded as beyond 'mathematicisation', and the continuing relevance and usefulness of his work confirms the value of his ideas. He generally worked in isolation, moving frequently from one subject to another. He lacked constructive collaboration with colleagues and, perhaps as a result, his work had great individuality but was also somewhat idiosyncratic. G. I. Taylor (1959) spoke of him as 'a very interesting and original character who seldom thought on the same lines as his contemporaries and often was not understood by them'. Just as for his work in meteorology, Richardson's mathematical studies of the causes of war were ahead of their time. In a letter to *Nature* (Richardson, 1951) he posed the question of whether an arms race must necessarily lead to warfare. Reviewing this work, his biographer (Ashford, 1985, p. 223) wrote 'Let us hope that before long history will show that an arms race can indeed end without fighting.' Just four years later the collapse of the Soviet Union brought the nuclear arms race to an abrupt end.[3]

Richardson's Quaker background and pacifist convictions profoundly influenced the course of his career. Late in his life, he wrote of the 'persistent influence of the Society of Friends, with its solemn emphasis on public and private duty'. Because of his pacifist principles, he resigned twice from the Met Office, first to face battlefield dangers in the Friends Ambulance Unit in France, and again when the Office came under the Air Ministry. He destroyed some of his research results to prevent their use for military purposes (Brunt, 1954) and even ceased meteorological research for a time: he published no papers in meteorology between 1930 and 1948. He retired early on a meagre pension to devote all his energies to peace studies. His work was misunderstood by many but his conviction and vision gave him courage to persist in the face of the indifference and occasional ridicule of his contemporaries.

Richardson made important contributions in several fields, the most significant being atmospheric diffusion, numerical analysis, quantitative psychology and the mathematical study of the causes of war. He is remembered by meteorologists through the Richardson Number, a fundamental quantity in turbulence theory, and for his extraordinary vision in formulating the process of numerical forecasting. The

---

[3] Stommel (1985) noted that the only purchaser of the book *Arms and Insecurity,* which Richardson was offering for sale on microfilm in 1948, was the Soviet Embassy in London!

approximate methods that he developed for the solution of differential equations are extensively used in the numerical treatment of physical problems.

Richardson's pioneering work in studying the mathematical basis of human conflict has led to the establishment of a large number of university departments devoted to this area. In the course of his peace studies, he digressed to consider the lengths of geographical borders and coastlines, and discovered the scaling properties such that the length increases as the unit of measurement is reduced. This work inspired Benoit Mandelbrot's development of the theory of fractals (Mandelbrot, 1982). In a tribute to Richardson shortly after his death, his wife Dorothy recalled that one of his sayings was 'Our job in life is to make things better for those who follow us. What happens to ourselves afterwards is not our concern.' Richardson had the privilege to make contributions to human advancement in several areas. The lasting value of his work is a testimony of his wish to serve his fellow man.

1.4 The origin of *Weather Prediction by Numerical Process*

Richardson first applied his approximate method for the solution of differential equations to investigate the stresses in masonry dams (Richardson, 1910), a problem on which he had earlier worked with the statistician Karl Pearson. But the method was completely general and he realised that it had potential for use in a wide range of problems. The idea of numerical weather prediction appears to have germinated in his mind for several years. In a letter to Pearson, dated 6 April 1907, he wrote in reference to the method that 'there should be applications to meteorology one would think' (Ashford, 1985, p. 25). This is the first inkling of his interest in the subject. In the preface to *WPNP* he wrote that the investigation of numerical prediction

grew out of a study of finite differences and first took shape in 1911 as the fantasy which is now relegated to Ch. 11/2. Serious attention to the problem was begun in 1913 at Eskdalemuir Observatory, with the permission and encouragement of Sir Napier Shaw, then Director of the Met Office, to whom I am greatly indebted for facilities, information and ideas.

The fantasy was that of a forecast factory, which we will discuss in detail in the final chapter. Richardson had had little or no previous experience of meteorology when he took up his position as Superintendent of the Observatory in what Gold (1954) described as 'the bleak and humid solitude of Eskdalemuir'. Perhaps it was this lack of formal training in the subject that enabled him to approach the problem of weather forecasting from such a breathtakingly original and unconventional angle. His plan was to express the physical principles that govern the behaviour of the atmosphere as a system of mathematical equations and to solve this system using his approximate finite difference method. The basic equations had already

Figure 1.5 Eskdalemuir Observatory in 1911. Office and Computing Room, where Richardson's dream began to take shape. (photograph from MC-1911)

been identified by Bjerknes (1904) but with the error noted above: the second law of thermodynamics was specified instead of conservation of water substance. The same error was repeated in Bjerknes' inaugural address at Leipzig (Bjerknes, 1914a). While this may seem a minor matter it proves that, while Bjerknes outlined a general philosophical approach, he did not attempt to formulate a detailed procedure, or algorithm, for applying his method. Indeed, he felt that such an approach was completely impractical. The complete system of fundamental equations was, for the first time, set down in a systematic way in Chapter 4 of *WPNP*. The equations had to be simplified, using the hydrostatic assumption, and transformed to render them amenable to approximate solution. Richardson also introduced a plethora of extra terms to account for various physical processes not considered by Bjerknes.

By the time of his resignation, in 1916, Richardson had completed the formulation of his scheme and had set down the details in the first draft of his book, then called *Weather Prediction by Arithmetic Finite Differences*. But he was not concerned merely with theoretical rigour and wished to include a fully worked example to demonstrate how the method could be put to use. This example

was worked out in France in the intervals of transporting wounded in 1916–1918. During the battle of Champagne in April 1917 the working copy was sent to the rear, where it became lost, to be re-discovered some months later under a heap of coal.

(WPNP, p. ix; *Dover Edn.,* p. xiii)

One may easily imagine Richardson's distress at this loss and the great relief that the re-discovery must have brought him.[4] It is a source of wonder that in the appalling conditions prevailing at the front he had the buoyancy of spirit to carry out one of the most remarkable and prodigious feats of calculation ever accomplished.

Richardson assumed that the state of the atmosphere at any point could be specified by seven numbers: pressure, temperature, density, water content and velocity components eastward, northward and upward. He formulated a description of atmospheric phenomena in terms of seven differential equations. To solve them, Richardson divided the atmosphere into discrete columns of extent 3° east–west and 200 km north–south, giving $120 \times 100 = 12\,000$ columns to cover the globe. Each of these columns was divided vertically into five cells. The values of the variables were given at the centre of each cell, and the differential equations were approximated by expressing them in finite difference form. The rates of change of the variables could then be calculated by arithmetical means. Richardson calculated the *initial changes in two columns* over central Europe, one for mass variables and one for winds. This was the extent of his 'forecast'.

How long did it take Richardson to make his forecast? It is generally believed that he took six weeks for the task but, given the volume of results presented on his 23 computing forms, it is difficult to understand how the work could have been expedited in so short a time. The question was discussed in Lynch (1993), which is reproduced in Appendix 4. The answer is contained in §11/2 of *WPNP*, but is expressed in a manner that has led to confusion. On page 219, under the heading 'The Speed and Organization of Computing', Richardson wrote:

It took me the best part of six weeks to draw up the computing forms and to work out the new distribution in two vertical columns for the first time. My office was a heap of hay in a cold rest billet. With practice the work of an average computer might go perhaps ten times faster. If the time-step were 3 hours, then 32 individuals could just compute two points so as to keep pace with the weather.

Could Richardson really have completed his task in six weeks? Given that 32 computers working at ten times his speed would require three hours for the job,[5] he himself must have taken some 960 hours – that is 40 days or 'the best part of six weeks' working flat-out at 24 hours a day! At a civilised 40-hour week the forecast would have extended over six months. It is more likely that Richardson spent perhaps ten hours per week at his chore and that it occupied him for about two years, the greater part of his stay in France.

In 1919, Richardson added an introductory example (*WPNP*, Chapter 2) in which he integrated a system equivalent to the linearised shallow water equations, starting

---

[4] In an obituary notice, Brunt (1954) stated that the manuscript was lost not once but twice.
[5] Richardson's 'computers' were made not of silicon but of flesh and blood.

from idealised initial conditions defined by a simple analytic formula. This was done at Benson where he had 'the good fortune to be able to discuss the hypotheses with Mr W. H. Dines'. The chapter ends with an acknowledgement to Dines for having read and criticised it. It seems probable that the inclusion of this example was suggested by Dines, who might have been more sensitive than Richardson to the difficulties that readers of *WPNP* would likely experience. The book was thoroughly revised in 1920–1 and was finally published by Cambridge University Press in 1922 at a price of 30 shillings (£1.50), the print run being 750 copies.

Richardson's book was certainly not a commercial success. Akira Kasahara has told me that he bought a copy from Cambridge University Press in 1955, more than thirty years after publication. The book was re-issued in 1965 as a Dover paperback and the 3 000 copies, priced at $2, about the same as the original hard-back edition, were sold out within a decade.[6] The Dover edition was identical to the original except for a six-page introduction by Sydney Chapman. Following its appearance, a retrospective appraisal of Richardson's work by George Platzman was published in the *Bulletin of the American Meteorological Society* (Platzman, 1967; 1968). This scholarly review has been of immense assistance in the preparation of the present work.

The initial response to *WPNP* was unremarkable and must have been disappointing to Richardson. The book was widely reviewed with generally favourable comments – Ashford (1985) includes a good coverage of reactions – but the impracticality of the method and the abysmal failure of the solitary sample forecast inevitably attracted adverse criticism. Napier Shaw, reviewing the book for *Nature*, wrote that Richardson 'presents to us a *magnum opus* on weather prediction'. However, in regard to the forecast, he observed that the wildest guess at the pressure change would not have been wider of the mark. More importantly for our purposes, he questioned Richardson's conclusion that wind observations were the real cause of the error, and also his dismissal of the geostrophic wind. Edgar W. Woolard, a meteorologist with the US Weather Bureau, wrote:

The book is an admirable study of an eminently important problem . . . a first attempt in this extraordinarily difficult and complex field . . . it indicates a line of attack on the problem, and invites further study with a view to improvement and extension. . . . It is sincerely to be hoped that the author will continue his excellent work along these lines, and that other investigators will be attracted to the field which he has opened up. The results cannot fail to be of direct practical importance as well as of immense scientific value.

However, other investigators were not attracted to the field, perhaps because the forecast failure acted as a deterrent, perhaps because the book was so difficult to read, with its encyclopedic but distracting range of topics. Alexander McAdie,

---

[6] *WPNP* has recently been reprinted by Cambridge University Press (2006).

Professor of Meteorology at Harvard, wrote 'It can have but a limited number of readers and will probably be quickly placed on a library shelf and allowed to rest undisturbed by most of those who purchase a copy' (McAdie, 1923). Indeed, this is essentially what happened to the book.

A most perceptive review by F. J. W. Whipple of the Met Office came closest to understanding Richardson's unrealistic forecast, postulating that rapidly-travelling waves contributed to its failure:

The trouble that he meets is that quite small discrepancies in the estimate of the strengths of the winds may lead to comparatively large errors in the computed changes of pressure. It is very doubtful whether sufficiently accurate results will ever be arrived at by the straight-forward application of the principle of conservation of matter. In nature any excess of air in one place originates waves which are propagated with the velocity of sound, and therefore much faster than ordinary meteorological phenomena.

One of the difficulties in the mathematical analysis of pressure changes on the Earth is that the great rapidity of these adjustments by the elasticity of the air has to be allowed for. The difficulty does not crop up explicitly in Mr Richardson's work, but it may contribute to the failure of his method when he comes to close quarters with a numerical problem.

The hydrostatic approximation used by Richardson eliminates vertically propagating sound waves, but gravity waves and also horizontally propagating sound waves (Lamb waves) are present as solutions of his equations. These do indeed travel 'much faster than ordinary meteorological phenomena'. Nowhere in his book does Richardson allude to this fact. Whipple appears to have had a far clearer understanding of the causes of Richardson's forecast catastrophe that did Richardson himself. The consideration of these causes is a central theme of the present work.

A humourist has observed that publishing a book of verse is like dropping a feather down the Grand Canyon and awaiting the echo. Richardson's work was not taken seriously and his book failed to have any significant impact on the practice of meteorology during the decades following its publication. But the echo finally arrived and continues to resound around the world to this day: Richardson's brilliant and prescient ideas are now universally recognised among meteorologists and his work is the foundation upon which modern forecasting is built.

1.5 Outline of the contents of *WPNP*

We will examine Richardson's numerical forecast in considerable detail in the chapters that follow. For now, it is useful to present a broad outline – a synoptic view – of his book. The chapter titles are given in Table 1.1. Chapter 1 is a summary of the contents of the book. Richardson's plan is to apply his finite difference method to the problem of weather forecasting. He had previously used both graphical and numerical methods for solving differential equations and had come to favour the latter:

Table 1.1 *Chapter titles of* Weather Prediction by
Numerical Process.

Chapter 1	Summary
Chapter 2	Introductory Example
Chapter 3	The Choice of Coordinate Differences
Chapter 4	The Fundamental Equations
Chapter 5	Finding the Vertical Velocity
Chapter 6	Special Treatment for the Stratosphere
Chapter 7	The Arrangement of Points and Instants
Chapter 8	Review of Operations in Sequence
Chapter 9	An Example Worked on Computing Forms
Chapter 10	Smoothing the Initial Data
Chapter 11	Some Remaining Problems
Chapter 12	Units and Notation

whereas Prof. Bjerknes mostly employs graphs, I have thought it better to procede by way of numerical tables. The reason for this is that a previous comparison of the two methods, in dealing with differential equations, had convinced me that the arithmetical procedure is the more exact and the more powerful in coping with otherwise awkward equations.

(WPNP, p. viii; Dover Edn., p. xii)

The fundamental idea is that the numerical values of atmospheric pressures, velocities, etc., are tabulated at certain latitudes, longitudes and heights so as to give a general description of the state of the atmosphere at an instant. The physical laws determine how these quantities change with time. The laws are used to formulate an arithmetical procedure, which, when applied to the numerical tables, yields the corresponding values after a brief interval of time, Δt. The process can be repeated so as to yield the state of the atmosphere after $2\Delta t$, $3\Delta t$, and so on, until the desired forecast length is reached.

In Chapter 2 the method of numerical integration is illustrated by application to a simple linear 'shallow-water' model. The step-by-step description of Richardson's method and calculations in this chapter is clear and explicit and is a splendid introduction to the process of numerical weather prediction. In contrast, the remainder of the book is heavy going, containing so much extraneous material that the central ideas are often obscured.

Chapter 3 describes the choice of co-ordinates and the discrete grid to be used. The choice is guided by (1) the scale of variation of atmospheric variables, (2) the errors due to replacing infinitesimal by finite differences, (3) the accuracy that is necessary to satisfy public requirements, (4) the cost, which increases with the number of points in space and time that have to be dealt with (*WPNP*, p. 16). Richardson considered the distribution of observing stations in the British Isles, which were separated, on average, by a distance of 130 km. Over the oceans, observations were 'scarce and irregular'. He concluded that a grid with 128 equally spaced meridians and 200 km in latitude would be a reasonable choice. In the vertical he chose five layers, or *conventional strata*, separated by horizontal surfaces at 2.0, 4.2, 7.2 and 11.8 km, corresponding approximately to the mean heights of the 800, 600, 400 and 200 hPa surfaces. The alternative of using isobaric co-ordinates was considered but dismissed. The time interval chosen by Richardson was six hours, but this corresponds to $2\Delta t$ for the leapfrog method of integration; in modern terms, we have $\Delta t = 3$ h. The cells of the horizontal grid were coloured alternately red and white, like the checkers of a chessboard. The grid was illustrated on the frontispiece of *WPNP*, reproduced in Fig. 1.6.[7]

The next three chapters, comprising half the book, are devoted to assembling a system of equations suitable for Richardson's purposes. In Chapter 4

the fundamental equations are collected from various sources, set in order and completed where necessary. Those for the atmosphere are then integrated with respect to height so as to make them apply to the mean values of the pressure, density, velocity, etc., in the several conventional strata.

As hydrostatic balance is assumed, there is no prognostic equation for the vertical velocity. Chapter 5 is devoted to the derivation of a diagnostic equation for this quantity. Platzman (1967) wrote that Richardson's vertical velocity equation 'is the principal, substantive contribution of the book to dynamic meteorology'. Chapter 6 considers the special measures that must be taken for the uppermost layer, the stratosphere, a region later described as 'a happy hunting-ground for meteorological theorists' (Richardson and Munday, 1926).

Chapter 7 gives details of the finite difference scheme, explaining the rationale for the choice of a staggered grid. Richardson considers several possible time-stepping techniques, including a fully implicit scheme, but opts for the simple leapfrog or 'step-over' method. Here can also be found a discussion of variable grid resolution and the special treatment of the polar caps. In Chapter 8 the forecasting 'algorithm' is presented in detail. It is carefully constructed so as to be, in Richardson's words, *lattice reproducing*; that is, where a quantity is known at a particular time and place, the algorithm enables its value at a later time to be calculated at the same place. The description of the method is sufficiently detailed and precise to enable a computer program based on it to be written, so that Richardson's results can be replicated (without the toil of two years' manual calculation).

Chapter 9 describes the celebrated trial forecast and its unfortunate results. The preparation of the initial data is outlined – the data are tabulated on p. 185 of *WPNP*.

[7] Richardson used 120 meridians, giving a 3° east–west distance, for his actual forecast, later realising that 128 meridians (or 2.8125°) would more conveniently facilitate sub-division near the poles.

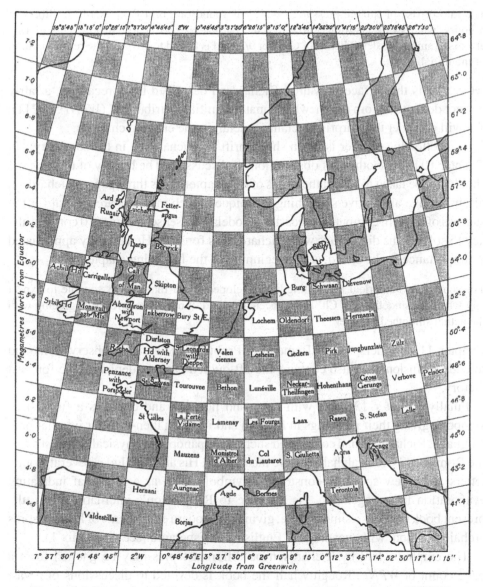

Figure 1.6 Richardson's idealised computational grid. (Frontispiece of *WPNP*)

The calculations themselves are presented on a set of 23 computing forms. These were completed manually: 'multiplications were mostly worked by a 25 centim slide rule' (*WPNP*, p. 186). The calculated changes in the primary variables over a six-hour period are compiled on page 211. It is characteristic of Richardson's whimsical sense of humour that, on the heading of this page, the word 'prediction' is enclosed in quotes; the results certainly cannot be taken literally. Richardson explains the chief result thus:

The rate of rise of surface pressure, $\partial p_G/\partial t$, is found on Form P_{XIII} as 145 millibars in 6 hours, whereas observations show that the barometer was nearly steady. This glaring error is examined in detail below in Chapter 9/3, and is traced to errors in the representation of the initial winds.

(Here, p_G is the surface pressure). Richardson described his forecast as 'a fairly correct deduction from a somewhat unnatural initial distribution' (*WPNP*, p. 211). We will consider this surprising claim in detail in the ensuing chapters.

The following chapter is given short shrift by Richardson in his summary: 'In Chapter 10 the smoothing of observations is discussed.' The brevity of this resumé should not be taken to reflect the status of the chapter. In its three pages, Richardson discusses five alternative smoothing techniques. Such methods are crucial for the success of modern computer forecasting models. In a sense, Chapter 10 contains the key to solving the difficulties with Richardson's forecast. He certainly appreciated its importance for he stated, at the beginning of the following chapter:

The scheme of numerical forecasting has developed so far that it is reasonable to expect that when the smoothing of Ch. 10 has been arranged, it may give forecasts agreeing with the actual smoothed weather.

Chapter 11 considers 'Some Remaining Problems' relating to observations and to eddy diffusion, and also contains the oft-quoted passage depicting the forecast factory.

Finally, Chapter 12 deals with units and notation and contains a full list of symbols, giving their meanings in English and in Ido, a then-popular international language. Richardson had considered such a vast panoply of physical processes that the Roman and Greek alphabets were inadequate. His array includes several Coptic letters and a few specially constructed symbols, such as a little leaf indicating evaporation from vegetation. As a tribute to Richardson's internationalism, the present book contains a similar table, giving the modern equivalents of Richardson's archaic notation, with meanings in English and Esperanto (see Appendix 1).

The emphasis laid by Richardson on different topics may be gauged from a page count of *WPNP*. Roughly half the book is devoted to discussions of a vast range of physical processes, some having a negligible effect on the forecast. The approximate *budget* in Table 1.2 is based on an examination of the contents of *WPNP* and on the earlier analyses of Platzman (1967) and Hollingsworth (1994). Due to the imprecision of the attribution process, the figures should be interpreted only in a qualitative sense.

The 23 computing forms on which the results of the forecast were presented, were designed and arranged in accordance with the systematic algorithmic procedure that Richardson had devised for calculating the solution of the equations. The completed

Table 1.2 *Page-count of* Weather Prediction by Numerical Process.

Dynamics	Momentum Equations	11		
	Vertical Velocity	10		
	The Stratosphere	24		
	Total Dynamics		45	
Numerics	Finite Differences	12		
	Numerical Algorithm	25		
	Total Numerics		37	
	Dynamics + Numerics			82
Physics	Clouds and Water	12		
	Energy and Entropy	8		
	Radiation	19		
	Turbulence	36		
	Surface, Soil, Sea	23		
	Total Physics			98
Miscellaneous	Summary	3		
	Initial Data	7		
	Analysis of Results	5		
	Smoothing	3		
	Forecast Factory	1		
	Computing Forms	23		
	Notation and Index	14		
	Total Miscellaneous			56
Total Pages				236

forms appear on pages 188–210 of *WPNP* so that the arithmetical work can be followed in great detail. Richardson arranged, at his own expense, for sets of blank forms to be printed to assist intrepid disciples to carry out experimental forecasts with whatever observational data were available. It is not known if these forms, which cost two shillings per set, were ever put to their intended use.[8]

The headings of the computing forms (see Table 1.3) indicate the scope of the computations. 'The forms are divided into two groups marked P and M according as the point on the map to which they refer is one where pressure *P* or momenta *M* are tabulated' (*WPNP*, p. 186). This arrangement of the computations is quite analogous to a modern spreadsheet program such as Excel, where the data are entered and the program calculates results according to prescribed rules. The first three forms contain input data and physical parameters. The forms may be classified as follows (Platzman, 1967):

---

[8] I am grateful to Oliver Ashford for providing me with a set of blank forms; they remain to be completed.

Table 1.3 *Headings of the 23 computing forms designed and used by Richardson.*
Copies were available separately from his book as Forms whereon to write the numerical calculations described in Weather Prediction by Numerical Process *by Lewis F. Richardson. Cambridge University Press, 1922. Price two shillings.*

Computing form	Title
P_I	Pressure, Temperature, Density, Water and Continuous Cloud
P_{II}	Gas constant. Thermal capacities. Entropy derivatives
P_{III}	Stability, Turbulence, Heterogeneity, Detached Cloud
P_{IV}	For Solar Radiation in the grouped ranges of wave-lengths known as *BANDS*
P_V	For Solar Radiation in the grouped ranges of wave-lengths known as *REMAINDER*
P_{VI}	For Radiation due to atmospheric and terrestrial temperature
P_{VII}	Evaporation at the interface
P_{VIII}	Fluxes of Heat at the interface
P_{IX}	For Temperature of Radiating Surface. Part I, Numerator of Ch. 8/2/15#20
P_X	For Temperature of Radiating Surface. Part II, Denominator of Ch. 8/2/15#20
P_{XI}	Diffusion produced by eddies. See Ch. 4/8. Ch. 8/2/13
P_{XII}	Summary of gains of entropy and of water, both per mass of atmosphere during δt
P_{XIII}	Divergence of horizontal momentum-per-area. Increase of pressure
P_{XIV}	Stratosphere. Vertical Velocity by Ch. 6/6#21. Temperature Change by Ch. 6/7/3#8
P_{XV}	For Vertical Velocity in general, by equation Ch. 8/2/23#1. Preliminary
P_{XVI}	For Vertical Velocity. Conclusion
P_{XVII}	For the transport of water and its increase in a fixed element of volume
P_{XVIII}	For water in soil $\frac{\partial w}{\partial t} = \cdots$, which is equation Ch. 4/10/2#5
P_{XIX}	For Temperature in soil. The equation is Ch. 4/10/2, namely $\frac{\partial \theta}{\partial t} = \cdots$
M_I	For Stresses due to Eddy Viscosity
M_{II}	Stratosphere. Horizontal velocities and special terms in dynamical equations
M_{III}	For the Dynamical Equation for the Eastward Component
M_{IV}	For the Dynamical Equation for the Northward Component

- *Hydrodynamic calculations (11 forms)*
  - Input data and physical parameters: P_I–P_{III}
  - Mass tendency and pressure tendency: P_{XIII}
  - Vertical velocity: P_{XIV}–P_{XVI}
  - Momentum tendency: M_I–M_{IV}

- *Thermodynamic and hydrologic calculations (12 forms)*
  - Radiation: $P_{IV}-P_{VI}$
  - Ground surface and sub-surface: $P_{VII}-P_X$, P_{XVIII}, P_{XIX}
  - Free air: P_{XI}, P_{XII}, P_{XVII}

The hydrodynamic calculations are by far the more important. In repeating the forecast we will omit the thermodynamic and hydrological calculations, which prove to have only a minor effect on the computed tendencies. The results on Form P_{XIII} are of particular interest and include the calculated surface pressure change of 145 hPa/6 h (the observed change in pressure over the period was less than one hPa).

Throughout his career, Richardson continued to consider the possibility of a second edition of *WPNP*. He maintained a file in which he kept material for this purpose and added to it from time to time, the last entry being in 1951. Platzman (1967) stressed the importance of this *Revision File* and discussed several items in it. The file contained an unbound copy of *WPNP*, on the sheets of which Richardson added numerous annotations. Interleaved among the printed pages were manuscript notes and correspondence relating to the book. In 1936, C. L. Godske, an assistant of Bjerknes, visited Richardson in Paisley to discuss the possibility of continuing his work using more modern observational data. Richardson gave him access to the *Revision File* and, after the visit, wrote to Cambridge University Press suggesting Godske as a suitable author if a second edition should be called for at a later time (Ashford, 1985, p. 157). After Richardson's death, the *Revision File* passed to Oliver Ashford who in 1967 deposited it in the archives of the Royal Meteorological Society. The file was misplaced, along with other Richardson papers, when the Society moved its headquarters from London to Bracknell in 1971. Ashford expressed a hope that 'perhaps it too will turn up some day 'under a heap of coal'.' The file serendipitously re-appeared around 2000 and Ashford wrote in a letter to *Weather* that 'there is still something of a mystery' about where the file had been (Ashford, 2001). The file has now been transferred to the National Meteorological Archive of the Met Office in Exeter. We will refer repeatedly in the sequel to this peripatetic file.

1.6 Preview of remaining chapters

The fundamental equations of motion are introduced in Chapter 2. The prognostic equations, which follow from the physical conservation laws, are presented and a number of diagnostic relationships necessary to complete the system are derived. In the case of small amplitude horizontal flow the equations assume a particularly simple form, reducing to the linear shallow-water equations or Laplace tidal equations.

These are discussed in Chapter 3, and an analysis of their normal mode solutions is presented. The numerical integration of the linear shallow-water equations is dealt with in Chapter 4. Richardson devoted a chapter of his book to this barotropic case, with the aim of verifying that his finite difference method could yield results of acceptable accuracy. We consider his use of geostrophic initial winds and show how the noise in his forecast may be filtered out.

The transformation of the full system of differential equations into algebraic form is undertaken in Chapter 5. This is done by the method of finite differences in which continuous variables are represented by their values at a discrete set of grid-points in space and time, and derivatives are approximated by differences between the values at adjacent points. The vertical stratification of the atmosphere is considered: the continuous variation is averaged out by integration through each of five layers and the equations for the mean values in each layer are derived. A complete system of equations suitable for numerical solution is thus obtained. A detailed step-by-step description of Richardson's solution procedure is given in this chapter.

The preparation of the initial conditions is described in Chapter 6. The sources of the initial data are discussed, and the transformations required to produce the needed initial values are outlined. There is also a brief description of the instruments used in 1910 in the making of these observations. In Chapter 7 the initial tendencies produced by the numerical model are presented. They are in excellent agreement with the values that Richardson obtained. The reasons for the small discrepancies are explained. The results are unrealistic: the reasons for this are analysed and we begin to consider ways around the difficulties.

The process of initialisation is discussed in Chapter 8. We review early attempts to define a balanced state for the initial data. The ideas of normal mode initialisation, filtered equations and the slow manifold are introduced by consideration of a particularly simple mechanical system, an elastic pendulum or 'swinging spring'. These concepts are examined in greater detail in the remaining sections of the chapter. Finally, the digital filter initialisation technique, which is later applied to Richardson's forecast, is presented.

In Chapter 9, we discuss the initialisation of Richardson's forecast. Richardson's discussion on smoothing the initial data is re-examined. When appropriate smoothing is applied to the initial data, using a simple digital filter, the initial tendency of surface pressure is reduced from the unrealistic 145 hPa/6 h to a reasonable value of less than 1 hPa/6 h. The forecast is shown to be in good agreement with the observed pressure change. The rates of change of temperature and wind are also realistic. To extend the forecast, smoothing in space is found to be necessary. The results of a 24-hour forecast with such smoothing are presented.

Chapter 10 considers the development of Numerical Weather Prediction (NWP) in the 1950s, when high-speed electronic computers first came into use. The first demonstration that computer forecasting might be practically feasible was carried out by the Princeton Group (Charney *et al.*, 1950). These pioneers were strongly impressed by Richardson's work as presented in his book. With the benefit of advances in understanding of atmospheric dynamics made since Richardson's time, they were able to devise means of avoiding the problems that had ruined his forecast. The Electronic Numerical Integrator and Computer (ENIAC) integrations are described in detail. There follows a description of the development of primitive equation modelling. The chapter concludes with a discussion of general circulation models and climate modelling.

The state of numerical weather prediction today is summarised in Chapter 11. The global observational system is reviewed, and methods of objectively analysing the data are described. The exponential growth in computational power is illustrated by considering the sequence of computers at the Met Office. To present the state of the art of Numerical Weather Prediction (NWP), the operations of the European Centre for Medium-Range Weather Forecasts (ECMWF) are reviewed. There follows a brief outline of current meso-scale modelling. The implications of chaos theory for atmospheric predictability are considered, and probabilistic forecasting using ensemble prediction systems is described.

In Chapter 12 we review Richardson's understanding of the causes of the failure of his forecast. His wonderful fantasy about a forecast factory is then re-visited. A parallel between this fantasy and modern massively parallel computers is drawn. Finally, we arrive at the conclusion that modern weather prediction systems provide a spectacular realisation of Richardson's dream of practical numerical weather forecasting.

Note added at proof stage

Willis and Hooke (2006) have recently reviewed the work of the great American meteorologist Cleveland Abbe (1838–1916). In his paper 'The physical basis of long-range weather forecasting', Abbe (1901) proposed a mathematical approach to forecasting by solution of the hydrodynamic and thermodynamic equations. Indeed, Abbe had been considering rational physical and mathematical approaches to forecasting for several decades. In 1905, Abbe acted as host to Bjerknes and arranged speaking engagements for his visit to the United States.

2

The fundamental equations

These equations contain the laws according to which subsequent atmospheric conditions develop from those that preceded them.

(Bjerknes, 1914a)

The behaviour of the atmosphere is governed by the fundamental principles of conservation of mass, energy and momentum. Conservation of mass ensures that matter cannot be created or destroyed. Conservation of energy implies that internal energy can be altered only by performance of work or by adding or removing heat. The law of motion states that the momentum can be changed only by a force. These principles, expressed in quantitative form, provide the framework for the study of atmospheric dynamics. The physical principles may be expressed mathematically in terms of differential equations. The prediction of the future development of atmospheric motion systems – the basis of weather forecasting – amounts to calculating the solution of these equations, given the state of the atmosphere at some initial time.

The idea of solving the equations to calculate future weather was propounded by Bjerknes in his famous 1904 manifesto but, although he outlined in principle how this might be done, he did not construct a detailed plan for implementing his programme or attempt to carry it through to a practical realisation. The first attempt to put this idea into practice was that of Richardson. In this chapter we will develop the equations used by Richardson, and begin to look at the way in which they might be solved.

Richardson was careful not to make any unnecessary approximations, and he took account of several physical processes that had the most marginal effect on his forecast. As we have seen, approximately half his book and more than half the computing forms were devoted to thermodynamic and hydrological processes and calculations that had only a minor influence on the outcome of his forecast. He included in his equations many terms that are negligible; with the benefit of hindsight, we can omit most of these.

2.1 Richardson's general circulation model

Since we are concerned primarily with the reasons for Richardson's catastrophic forecast failure, which can be treated from a purely dynamical viewpoint, it is feasible and convenient to disregard all diabatic processes. In the sequel, we will ignore all the effects of moisture and thermal forcing, and consider the adiabatic evolution of a dry atmosphere. However, a full appreciation of Richardson's work requires at least a brief examination of the wide variety of physical processes that he discussed. Many of the physical phenomena relevant to the atmosphere were considered for the first time by Richardson. He constructed the basis for what was, in effect, a comprehensive physics parameterisation package. If all the factors treated by him were included, one would have a comprehensive model of the circulation of the atmosphere, or, in modern terms, a general circulation model (GCM).

Richardson's description of physical processes was comprehensive and he contributed much that was original. His quantitative formulations of physical processes were based on the best field data available to him. Indeed, he carried out several innovative field experiments himself to measure a range of physical parameters. Chapter 4 of *WPNP*, entitled *The Fundamental Equations*, is 94 pages long and, in addition to the basic dynamical equations, contains detailed discussions of all major physical processes. The role of water in all its phases, and the thermodynamic consequences of phase changes, are discussed. Clouds and precipitation processes are considered: 'In order to save labour I have supposed there to be a sharp distinction between rain which falls and clouds which float. Actually there is a gradual transition' (*WPNP*, page 44). Short-wave solar radiation and long-wave terrestrial radiation and their interaction with the atmosphere are treated. Richardson notes the lack of observational data at wavelengths exceeding 15μm, remarking that, until such data is available, meteorologists 'must carry on business on premises which are, so to speak, in the hands of the builders' (*WPNP*, page 49). He introduces the concept of a 'parcel' of air (*WPNP*, page 50), an idea that has gained great popularity and utility.

The longest section of Chapter 4 deals with eddy motions. In this section, Richardson presents his famous rhyme, 'Big whirls have little whirls that feed on their velocity, and little whirls have lesser whirls and so on to viscosity', that beautifully encapsulates the essence of the turbulent energy cascade. He made fundamental contributions to turbulence theory and his writings here can still be read with profit. The dense style is occasionally lightened by a whimsical touch as, when discussing the tendency of turbulence to increase diversity, he writes 'This one can believe without the aid of mathematics, after watching the process of stirring together water and lime-juice' (*WPNP*, page 101). Finally, Richardson discusses the interaction between the atmosphere and the sea and land surfaces beneath it. He suggests that climatological sea temperatures may suffice, but also discusses how the sea surface temperature might be predicted. He considers heat and moisture transports within the soil and discusses at some length the influence of vegetation: 'Leaves, when present, exert a paramount influence on the interchanges of moisture and heat' (*WPNP*, page 111).[1] Clearly, Richardson is thinking far beyond short-range forecasting here, and has entered the realm of climate modelling.

The scope of Richardson's treatment of physical processes may be appreciated by examining the running heads of his Chapter 4; these are presented in Table 2.1. A full appraisal of his work in this area would entail a more intensive investigation than can be undertaken here:

We will not have a true appreciation of Richardson's achievement in atmospheric modelling until his suite of physical parameterizations is implemented in his own or some other dynamical framework, and its performance validated by comparison with the best current formulations of physical processes.   *(Hollingsworth, 1994)*

Let us hope that, before too long, someone with the requisite expertise, energy and enthusiasm will undertake this task.

2.2 The basic equations

We will set out the basic equations as commonly used today, and then convert them to the form used by Richardson. Some of Richardson's notation is archaic and the modern equivalents will be used (a Table of notation used in this book appears in Appendix A; a full list of Richardson's notation may be found in Chapter XII of *WPNP*).

2.2.1 The exact equations

The basic principles of motion are embodied in Newton's second law: the rate of change of momentum is equal to the applied force. Euler formulated the equations of motion as they apply to the flow of a continuous fluid. Laplace, in his study of tides, took account of the dynamical effects of rotation, and derived the equations of motion in a frame of reference spinning with the Earth. Later, Gustave Gaspard Coriolis again showed how the equations must be modified to account for a rotating frame of reference. Coriolis was studying rotating hydraulic machinery; he did not consider the geophysical consequences of rotation, but he elucidated its dynamical effects.[2] Navier and Stokes derived the form of the additional terms required to

---

[1] Richardson even allows (on Form P_{VIII}) for the insolation due to a layer of dead leaves on the ground – in May!

[2] 'The term "Coriolis acceleration" ... is frequently used by oceanographers and meteorologists without appreciation that it was first introduced by Laplace before Coriolis was born.' (Cartwright, 1999)

Table 2.1 *Running heads,* Weather Prediction by Numerical
Process, *Chapter 4.*

allow for the effects of frictional forces. The full equations of motion are now called the Navier–Stokes equations.

The forces influencing the motion of the atmosphere are those due to pressure gradients, gravity and friction and the apparent deflective forces due to rotation. If we consider a unit mass of air, the equation of motion may be written

$$\frac{d\mathbf{U}}{dt} = -2\mathbf{\Omega} \times \mathbf{U} - \frac{1}{\rho}\nabla p + \mathbf{F} + \mathbf{g}^* - \mathbf{\Omega} \times (\mathbf{\Omega} \times \mathbf{r}). \tag{2.1}$$

The dependent variables here are the velocity $\mathbf{U}$ relative to the rotating Earth, the pressure p and the density ρ. The independent variables are the radius vector $\mathbf{r}$ from the Earth's centre and the time t. The frictional force is denoted $\mathbf{F}$. This equation is valid for a frame of reference fixed to the Earth and rotating with angular velocity $\mathbf{\Omega}$. The effects of rotation are accounted for by the Coriolis force $-2\mathbf{\Omega} \times \mathbf{U}$ and a centrifugal term $-\mathbf{\Omega} \times (\mathbf{\Omega} \times \mathbf{r})$. Since the latter depends only on position, we may combine it with the true gravity $\mathbf{g}^*$ to produce an apparent gravitational acceleration

$$\mathbf{g} = \mathbf{g}^* - \mathbf{\Omega} \times (\mathbf{\Omega} \times \mathbf{r}).$$

This composite force is the quantity actually measured at the Earth's surface. The magnitude g of $\mathbf{g}$ varies by only a few per cent below 100 km (although, in his meticulous fashion, Richardson allowed for variations in g; see §2.2.3 below).

Equation (2.1) is effectively in the form set down by the American meteorologist William Ferrel in about 1860 (of course, his notation was different). Ferrel's derivation of the equations as they apply in a rotating frame fixed to the Earth was independent of the work of Coriolis, and sprang directly from the *Méchanique Céleste* of Laplace. Ferrel (1859) was the first to present a comprehensive treatment of the geophysical implications of the deflecting force due to rotation, and to deduce its consequences for the general circulation of the atmosphere and oceans (Kutzbach, 1979). He gave a quantitative description of the geostrophic wind and an account of the thermal wind relation. Ferrel considered the possibility of deriving a mathematical expression for the general circulation but felt that this could not be done because the frictional terms were inadequately known. Lorenz (1967) observed that:

It was a great loss to nineteenth-century meteorology that the man who introduced the equations of motion never saw fit to seek a complete solution of them.

The indestructability of mass is expressed in terms of the continuity equation

$$\frac{d\rho}{dt} + \rho\nabla \cdot \mathbf{U} = 0.$$

Here, as in (2.1), the time derivative is the material derivative following the flow:

$$\frac{d(\ )}{dt} = \frac{\partial (\ )}{\partial t} + \mathbf{U} \cdot \nabla (\ ).$$

Using this, the continuity equation may be written in *flux form*:

$$\frac{\partial \rho}{\partial t} + \nabla \cdot \rho \mathbf{U} = 0. \tag{2.2}$$

The principle of conservation of energy, which is the basis of thermodynamics, was formulated in the nineteenth century by Clausius, Helmholtz, Joule and Kelvin amongst others. In the context of atmospheric dynamics, it states that the heat energy added to a parcel of air may increase its internal energy or induce it to do work by expansion, and that the sum of these is equal to the energy supplied. Thus, the first law of thermodynamics gives the change of the internal energy ($c_v T$) of a unit mass of air, in terms of work done and heat energy supplied:

$$c_v \frac{dT}{dt} + p \frac{d}{dt}\left(\frac{1}{\rho}\right) = \dot{Q}, \tag{2.3}$$

where T is the temperature, c_v the specific heat at constant volume and $\dot{Q}$ the heating rate.

Finally, we treat the atmosphere as an ideal gas, obeying Boyle's Law and Charles' Law, so that the equation of state is

$$p = \Re \rho T \tag{2.4}$$

where $\Re$ is the gas constant for dry air ($\Re = c_p - c_v$ is the difference of the specific heats at constant pressure and volume).

Equations (2.1)–(2.4) comprise a complete set of equations for the dependent variables ($\mathbf{U}, p, \rho, T$), provided we can specify the energy sources and sinks represented by $\dot{Q}$ and $\mathbf{F}$. The ultimate source of all atmospheric motion is the energy radiated by the Sun. The source term $\dot{Q}$ is thus the dominant factor in determining the dynamics and climate of the atmosphere. However, for timescales of a day or so, a reasonable approximation to reality is obtained when diabatic forcing is ignored. As this hugely simplifies the problem, we will disregard all radiative and other diabatic processes by setting $\dot{Q} = 0$. We may also ignore the frictional drag $\mathbf{F}$, which has a relatively small impact in the free atmosphere. Finally, we have omitted all consideration of moisture, which is of paramount importance in the real atmosphere. Richardson devoted considerable attention to such matters but it transpired that they did not affect his final results in any major way. With these simplifying assumptions, the system of equations is complete.

2.2.2 The primitive equations

A predominant feature of the atmosphere is that the gravitational force is almost exactly cancelled by the vertical pressure gradient force. The assumption that this balance holds exactly is called the hydrostatic approximation, expressed as

$$\frac{\partial p}{\partial z} + g\rho = 0. \tag{2.5}$$

This fundamental approximation was made by Richardson. It was an essential step, necessitated by the lack of observations of vertical velocity, w. Since there is no longer an equation for dw/dt, another means of deducing the vertical velocity is required. We return to this in §2.3 below. Richardson's adoption of the hydrostatic approximation was influenced by the work of Bjerknes (*WPNP*, p. viii; Dover Edition, p. xii).

Following Richardson, we now simplify the exact equations by introducing the hydrostatic approximation, together with assumption that the atmosphere is a thin layer on the Earth's surface (the shallow atmosphere assumption). We also make the 'traditional approximation' of neglecting the Coriolis terms involving the vertical velocity. Richardson included the Coriolis terms proportional to $\cos \phi$. However, it is now known that, for a shallow atmosphere, it is dynamically inconsistent to include these terms. They must be omitted in order to maintain the conservation of angular momentum if the shallowness assumption is adopted (Phillips, 1966).

The result of the above approximations is the set of equations known as the primitive equations. They may be found in standard works on dynamic meteorology (e.g., Lorenz, 1967; Phillips, 1973; Holton, 2004). Lorenz presents a detailed derivation of the primitive equations, carefully clarifying each approximation that he makes. He stresses the requirement that appropriate energy and angular momentum principles should hold for any approximate system.

The horizontal equations of motion, in co-ordinate form, become:

$$\frac{\partial u}{\partial t} + u\frac{\partial u}{\partial x} + v\frac{\partial u}{\partial y} + w\frac{\partial u}{\partial z} - \left(f + \frac{u \tan \phi}{a} \right) v + \frac{1}{\rho}\frac{\partial p}{\partial x} = 0 \tag{2.6}$$

$$\frac{\partial v}{\partial t} + u\frac{\partial v}{\partial x} + v\frac{\partial v}{\partial y} + w\frac{\partial v}{\partial z} + \left(f + \frac{u \tan \phi}{a} \right) u + \frac{1}{\rho}\frac{\partial p}{\partial y} = 0. \tag{2.7}$$

The Earth's radius is a, its angular velocity is Ω and $f = 2\Omega \sin \phi$ is the Coriolis parameter. Distances eastward and northward on the globe are represented by x and y so that

$$\frac{\partial}{\partial x} = \frac{1}{a \cos \phi}\frac{\partial}{\partial \lambda} \quad \text{and} \quad \frac{\partial}{\partial y} = \frac{1}{a}\frac{\partial}{\partial \phi},$$

where λ and ϕ are the longitude and latitude. The momentum equation in vector form is

$$\left(\frac{d\mathbf{V}}{dt}\right)_H + f\mathbf{k} \times \mathbf{V} + \frac{1}{\rho}\nabla p = 0, \tag{2.8}$$

where $\mathbf{V} = (u, v)$ is the horizontal velocity, ∇ is the horizontal gradient operator with $1/a$ instead of $1/r$, $(d/dt)_H$ is the horizontal component of the total time-derivative and $\mathbf{k}$ is a vertical unit vector.

The continuity equation (2.2) becomes

$$\frac{\partial\rho}{\partial t} + \frac{\partial\rho u}{\partial x} + \frac{\partial\rho v}{\partial y} - \frac{\rho v \tan\phi}{a} + \frac{\partial\rho w}{\partial z} = 0 \tag{2.9}$$

(a negligibly small term $2\rho w/a$ has been dropped; naturally, Richardson retained it). In combination with this equation, the horizontal equations of motion may be written in flux form:

$$\frac{\partial\rho u}{\partial t} + \frac{\partial\rho u^2}{\partial x} + \frac{\partial\rho uv}{\partial y} + \frac{\partial\rho uw}{\partial z} - \left(f + \frac{2u\tan\phi}{a}\right)\rho v + \frac{\partial p}{\partial x} = 0$$

$$\frac{\partial\rho v}{\partial t} + \frac{\partial\rho vu}{\partial x} + \frac{\partial\rho v^2}{\partial y} + \frac{\partial\rho vw}{\partial z} + f\rho u + \frac{(\rho u^2 - \rho v^2)\tan\phi}{a} + \frac{\partial p}{\partial y} = 0.$$

The adiabatic thermodynamic equation in co-ordinate form is

$$c_v\left(\frac{\partial T}{\partial t} + u\frac{\partial T}{\partial x} + v\frac{\partial T}{\partial y} + w\frac{\partial T}{\partial z}\right) - \frac{p}{\rho^2}\left(\frac{\partial\rho}{\partial t} + u\frac{\partial\rho}{\partial x} + v\frac{\partial\rho}{\partial y} + w\frac{\partial\rho}{\partial z}\right) = 0. \tag{2.10}$$

Using the equation of state (2.4), this equation may be written

$$\frac{1}{\gamma p}\left(\frac{\partial p}{\partial t} + u\frac{\partial p}{\partial x} + v\frac{\partial p}{\partial y} + w\frac{\partial p}{\partial z}\right) - \frac{1}{\rho}\left(\frac{\partial\rho}{\partial t} + u\frac{\partial\rho}{\partial x} + v\frac{\partial\rho}{\partial y} + w\frac{\partial\rho}{\partial z}\right) = 0 \tag{2.11}$$

where $\gamma = c_p/c_v$ is the ratio of specific heats. The potential temperature is defined by $\theta = T(p/p_0)^{-\kappa}$, where $\kappa = \mathfrak{R}/c_p$ and $p_0 = 1000\,\text{hPa}$. The adiabatic thermodynamic equation expresses conservation of potential temperature

$$\frac{d\theta}{dt} = \left(\frac{\partial\theta}{\partial t} + u\frac{\partial\theta}{\partial x} + v\frac{\partial\theta}{\partial y} + w\frac{\partial\theta}{\partial z}\right) = 0. \tag{2.12}$$

Defining the entropy per unit mass by $S = c_p\log\theta$, we see that entropy is conserved in adiabatic flow.

The complete system of equations

For convenience, we assemble the complete system of equations:

$$\frac{\partial \rho u}{\partial t} + \frac{\partial \rho u^2}{\partial x} + \frac{\partial \rho u v}{\partial y} + \frac{\partial \rho u w}{\partial z} - \left(f + \frac{2u \tan \phi}{a} \right) \rho v + \frac{\partial p}{\partial x} = 0 \qquad \text{(Ch. 4/4#3)}$$

$$\frac{\partial \rho v}{\partial t} + \frac{\partial \rho v u}{\partial x} + \frac{\partial \rho v^2}{\partial y} + \frac{\partial \rho v w}{\partial z} + f \rho u + \frac{\left(\rho u^2 - \rho v^2 \right) \tan \phi}{a} + \frac{\partial p}{\partial y} = 0 \qquad \text{(Ch. 4/4#4)}$$

$$\frac{\partial p}{\partial z} + g \rho = 0 \qquad \text{(Ch. 4/2#2)}$$

$$\frac{\partial \rho}{\partial t} + \frac{\partial \rho u}{\partial x} + \frac{\partial \rho v}{\partial y} - \frac{\rho v \tan \phi}{a} + \frac{\partial \rho w}{\partial z} = 0 \qquad \text{(Ch. 4/4#6)}$$

$$c_v \left(\frac{\partial T}{\partial t} + u \frac{\partial T}{\partial x} + v \frac{\partial T}{\partial y} + w \frac{\partial T}{\partial z} \right) - \frac{p}{\rho^2} \left(\frac{\partial \rho}{\partial t} + u \frac{\partial \rho}{\partial x} + v \frac{\partial \rho}{\partial y} + w \frac{\partial \rho}{\partial z} \right) = 0 \qquad \text{(Ch. 4/5/0#18)}$$

$$p = \Re \rho T \qquad \text{(Ch. 4/1#1)}$$

Here, for example, the marginal number (Ch. 4/5/0#18) indicates a correspondence with equation (18) in §5/0 of Chapter 4 of *WPNP*. As we noted, Richardson made the shallow atmosphere assumption, replacing $r = a + z$ by a, but not the 'traditional approximation' of omitting the horizontal component of the Coriolis force and some small metric terms. Otherwise the equations assembled here are equivalent to his, with one curious exception, variations in the gravitational attraction, which we examine next.

2.2.3 *Variations of gravity with height and latitude*

The Earth is very close to a perfect sphere. The centrifugal force due to rotation has distorted the planet into an oblate spheroid, but the eccentricity is very small and deviations from sphericity are generally ignored; the difference between equatorial and polar radii is only about 20 km (about 0.3 per cent). The centrifugal term is combined with the true gravity to produce an 'apparent gravity'. This force is everywhere perpendicular to the spheroid and defines a vertical direction on its surface so that the apparent gravity has no horizontal component on this surface.

We can write the Newtonian gravitational potential as $\Phi_e = -g_0 a^2 / r$, where $g_0 = 9.80665 \text{ m s}^{-2}$ is the standard value of gravitational acceleration.[3] We also

---

[3] Variations of surface g with latitude are ignored, as we are primarily interested in the *horizontal component* of gravity, not the amplitude of the vertical component.

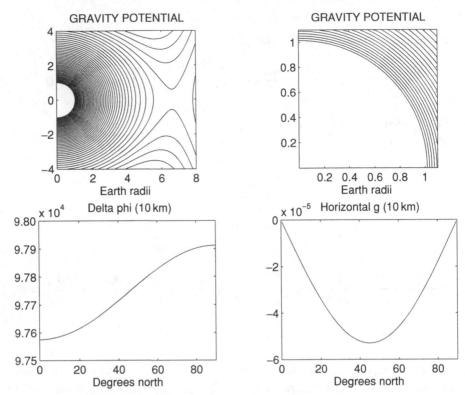

Figure 2.1 Potential of apparent gravity. Top left: broad field, to eight Earth radii. Top right: close-up view. Bottom left: $\Delta\Phi = \Phi(10 \text{ km}) - \Phi(\text{mean sea level})$. Bottom right: horizontal component $g_H = -\partial\Delta\Phi/\partial y$ of **g** at 10 km.

note that the centrifugal acceleration $-\mathbf{\Omega} \times (\mathbf{\Omega} \times \mathbf{r})$ is the gradient of the potential $\Phi_c = -\frac{1}{2}\Omega^2 R^2$, where R is the distance from the Earth's axis. Thus, the potential of apparent gravity is

$$\Phi = \Phi_e + \Phi_c = -g_0\frac{a^2}{r} - \frac{1}{2}\Omega^2 R^2.$$

This is plotted in Fig. 2.1 (top left panel). Near the Earth, the potential surfaces are approximately spherical (actually, spheroidal with minute eccentricity). Further out, they are greatly distorted. The equilibrium point over the equator at about seven Earth radii is valuable real estate, the location for geostationary satellites. Zooming in (top right panel), we see no evidence of asymmetry near the Earth. However, examining the difference $\Delta\Phi$ between the potential at 10 km and the value at the surface as a function of latitude (lower left panel), there is a perceptable gradient. This implies a horizontal component of gravity $g_H = -\partial\Delta\Phi/\partial y$ (lower right panel) directed towards the equator. Its maximum amplitude is about $5 \times 10^{-5} \text{ m s}^{-2}$. Richardson was very cautious about making unnecessary assumptions. He remarks (*WPNP*, page 22) 'As the arithmetical method allows us to take account of the

terms which are usually neglected, many of these terms have been included.'
In particular, he allows for small variations in the gravitational acceleration. He
includes a horizontal component of **g** which he denotes by g_N and tabulates as a
function of latitude and height. Thus, his northward momentum equation can be
written

$$\frac{\partial \rho v}{\partial t} = \rho g_N + \frac{\partial p}{\partial y} + \{\text{terms involving velocities}\} \qquad (2.13)$$

(see his Chapter 4/4#4 on page 31 of *WPNP*). Since g_N vanishes at the Earth's
surface, there will be no acceleration there if the velocity is initially zero and p
is independent of y. However, above the surface a meridional pressure gradient
is required to balance g_N if a steady state is to be maintained. A mean value of
$5 \times 10^{-5}\,\text{m s}^{-2}$ would require for balance an equator-to-pole height difference
of only about 30 m, a small fraction of the observed mean height difference at
250 hPa (annual mean about 1200 m). Nevertheless, it is a systematic effect.
Richardson's tabulated values of g_N include variations in true gravity due to the
Earth's eccentricity (as usual, he sought to use the most accurate values available
to him). While these values differ somewhat from the values of $g_H = -\partial \Delta \Phi / \partial y$
defined above, the qualitative arguments still apply.

The term ρg_N in (2.13) amounts to about 17 per cent of the total acceleration
for the uppermost layer in Richardson's scheme (Form M_{IV}). This effect is usually
ignored in current atmospheric models. The 'horizontal' component of gravity can
be removed by choosing the gravitational potential Φ as a 'vertical' co-ordinate
in place of z. This attractive alternative was briefly considered by Richardson and
described by him as 'a rather tempting one', but he resisted the temptation to use
it. Geopotential co-ordinates are discussed by Phillips (1973). With an appropriate
origin, Φ/g_0 is very close in value to z for the lower atmosphere.

2.3 The vertical velocity equation

If progress is to be possible, it can only be by eliminating the vertical velocity.

(*WPNP*, page 115)

The vertical component of velocity in the atmosphere is typically two or three orders
of magnitude smaller than the horizontal components. It is difficult to measure w and
in general no observations of this variable are available. In particular, Richardson
had no such observations for 0700 UTC on 20 May 1910, the initial time chosen
for his forecast. Moreover, even if he had had such observations, he recognised the
practical impossibility of computing the tendency $\partial w / \partial t$ which would have to be
calculated as a tiny residual term in the vertical dynamical equation.

Richardson acknowledged the influence of Vilhelm Bjerknes' publications
Statics and *Kinematics* (Bjerknes *et al.*, 1910, 1911) on his work. In his preface

(*WPNP*, p. viii; Dover Edition, p. xii) Richardson states that his choice of 'conventional strata', his use of specific momentum rather than velocity, his method of calculating vertical motion at ground level and his adoption of the hydrostatic approximation are all in accordance with Bjerknes' ideas.

The hydrostatic equation results from neglecting the vertical acceleration, and other small terms, in the vertical dynamical equation. But this precludes the possibility of calculating the acceleration $\partial w/\partial t$ directly. It was a stroke of genius for Richardson not only to realise the need to evaluate w diagnostically from the other fields but also to construct a magnificent mathematical equation to achieve this.

2.3.1 The tendency equation

An equation for the pressure tendency can be derived from the hydrostatic equation and continuity equation. The integrated form of the hydrostatic equation is

$$p = \int_{z}^{\infty} g\rho \, dz. \tag{2.14}$$

This is a mathematical statement of the physical assumption that the pressure at any point is determined by the weight of air above it. The continuity equation (2.9) may be written

$$\frac{\partial \rho}{\partial t} + \nabla \cdot \rho \mathbf{V} + \frac{\partial \rho w}{\partial z} = 0$$

where $\mathbf{V}$ is the horizontal velocity and $\nabla \cdot (\ )$ the horizontal divergence operator. Taking the time-derivative of (2.14), noting that the limits of integration are independent of time, and using the continuity equation, we obtain the following equation:

$$\frac{\partial p}{\partial t} = g\rho w - \int_{z}^{\infty} g\nabla \cdot \rho \mathbf{V} \, dz \tag{2.15}$$

(the boundary condition $\rho w \to 0$ as $z \to \infty$ has been used). This equation may be solved for the pressure tendency if the vertical velocity is known. In particular, if the lower limit is taken at $z = 0$ and the bottom boundary is assumed to be flat so that $w = 0$ there, we get

$$\frac{\partial p_S}{\partial t} = -\int_{0}^{\infty} g\nabla \cdot \rho \mathbf{V} \, dz, \tag{2.16}$$

sometimes called the *tendency equation*. This equation was discussed by Margules (1904) who recognised the impracticality of using it directly to forecast changes in pressure. He showed that tiny errors in the wind fields can result in spuriously large values for convergence of momentum and correspondingly unrealistic pressure tendency values (see §7.5). This was discovered, to his cost, by Richardson with the result that he obtained an unreasonable value for the pressure change; we will discuss this at length below.

2.3.2 Richardson's equation for vertical velocity

To construct Richardson's w-equation we eliminate the time dependency between the continuity equation and the thermodynamic equation using the hydrostatic equation. Recall that the thermodynamic equation can be written in the form

$$\frac{1}{\gamma p}\left(\frac{\partial p}{\partial t} + \mathbf{V}\cdot\nabla p + w\frac{\partial p}{\partial z}\right) - \frac{1}{\rho}\frac{d\rho}{dt} = 0, \tag{2.17}$$

and that one of the various forms of the continuity equation is

$$\frac{1}{\rho}\frac{d\rho}{dt} + \left(\nabla\cdot\mathbf{V} + \frac{\partial w}{\partial z}\right) = 0. \tag{2.18}$$

We can eliminate the pressure tendency in (2.17) using (2.15) and the density term by means of the continuity equation (2.18) to get

$$\frac{1}{\gamma p}\left(-\int_z^\infty g\nabla\cdot\rho\mathbf{V}\,dz + \mathbf{V}\cdot\nabla p\right) + \left(\nabla\cdot\mathbf{V} + \frac{\partial w}{\partial z}\right) = 0.$$

Expanding the integrand, using the hydrostatic equation again and rearranging, we get

$$\frac{\partial w}{\partial z} = -\nabla\cdot\mathbf{V} + \frac{1}{\gamma p}\int_z^\infty\left(g\rho\nabla\cdot\mathbf{V} - \frac{\partial\mathbf{V}}{\partial z}\cdot\nabla p\right)dz. \tag{2.19}$$

Since the upper limit of the integral is infinite, it is convenient to use pressure as the independent variable in the integral; this is done by using the hydrostatic equation once more, yielding the result:

$$\frac{\partial w}{\partial z} = -\nabla\cdot\mathbf{V} + \frac{1}{\gamma p}\int_0^p\left(\nabla\cdot\mathbf{V} - \frac{\partial\mathbf{V}}{\partial p}\cdot\nabla p\right)dp. \tag{2.20}$$

This corresponds to (9) on page 124 of *WPNP*, save that we have omitted the effects of moisture and diabatic forcing, which were included by Richardson. Following Eliassen (1949) we call (2.19), or (2.20), *Richardson's Equation*.

The solution of (2.20) for w is straightforward. The gradient $\partial w/\partial z$ is calculated for each layer, working downwards from the stratosphere since the integral vanishes at $p = 0$. Then w may be calculated at the interface of each layer, working upwards, once it is known at the Earth's surface. Richardson followed Bjerknes in taking the surface value

$$w_S = (\mathbf{V}\cdot\nabla h)_S,$$

where h is the surface elevation. This is equivalent to the kinematic condition that the ground is impervious to the wind. However, Richardson does not state how he evaluates V_S, the horizontal wind at the surface; he merely states (*WPNP*, p. 178) that it has to be estimated from the statistics of its relation to the horizontal wind in the lowest layer. In repeating his forecast we have assumed a simple relationship

$$V_S = kV_5 \qquad (2.21)$$

where V_5 is the velocity for the lowest layer, and have fixed the constant of proportionality arbitrarily by chosing the value $k = 0.2$.

Richardson described a complicated method of calculating the vertical derivative of w in the stratosphere (his equation (21), *WPNP*, p. 138). This was required because he used a differentiated form of the vertical velocity equation (his (1) on page 178) and needed a boundary condition to integrate it. He alluded to this on page 119 of *WPNP*: 'As a matter of fact, the differential form was first derived directly from ... [the hydrostatic equation] and was employed throughout the example of Ch. 9; but a constant of integration kept on appearing inconveniently in places where it could not be determined. This "hysterical manifestation" was eventually traced to the suppression of the limits of integration which are now explicit in equation ... [(2.20)]'. We take the simpler path of using (2.20) directly so that only one boundary condition (w at the ground) is required. The vertical velocities thus obtained will be compared to the values in Richardson's computing form P_{XVI}, and reasonable agreement shown.

The vertical velocity equation was a major contribution by Richardson to dynamic meteorology. In recognising its essential role in his forecast scheme he observed (*WPNP*, p. 178) that 'it might be called the keystone of the whole system, as so many other equations remain incomplete until the vertical velocity has been inserted'. In a hand-written note in the *Revision File*, Richardson draws an analogy between his forecasting algorithm and the workings of a motor-car engine. The 'enormous vertical velocity equation ... corresponds to the connecting rod for transmitting the power from the cylinders to the wheels. It is the sort of connecting rod that Heath Robinson[4] would delight to draw. And yet any connecting rod, even an ungainly one, is better than no connecting rod at all. And I am afraid there are to be found theories which omit this necessary link.'

2.4 Temperature in the stratosphere

Richardson devoted a full chapter of 24 pages to the stratosphere. We will not discuss the bulk of this, but we must consider the means by which the temperature of the

---

[4] W. Heath Robinson (1872–1944) was a British cartoonist and illustrator who delighted in drawing outlandish and ingeneous mechanical contraptions. His autobiography, *My Line of Life*, was published in 1938.

uppermost layer is forecast. For, in the scheme adopted by Richardson, the vertical integral of pressure through the stratospheric layer depends on the temperature so that prediction of the latter is essential to ensure a 'lattice-reproducing' scheme – that is, an algorithm which, starting with a set of variables at one instant, produces the corresponding set at a later instant.

Richardson calculated the change in stratospheric temperature using two different equations, his elaborate equation (8) on page 147 of *WPNP* and a much simpler equation corresponding to (14) on page 143. The resulting temperature tendencies, given in his computing form P$_{XIV}$ on page 201, were $9.1 \times 10^{-4}\,\mathrm{K\,s^{-1}}$ for the elaborate equation and $9.2 \times 10^{-4}\,\mathrm{K\,s^{-1}}$ for the simpler. In view of this close agreement, we will confine attention to the simpler alternative, which will now be derived.

The basic assumption for the stratosphere is that of vertical isothermy,

$$\frac{\partial T}{\partial z} = 0. \tag{2.22}$$

Since the effects of radiation are neglected here, the entropy $S = c_p \log \theta$ is conserved following the flow:

$$\frac{\partial S}{\partial t} + \mathbf{V} \cdot \nabla S + w \frac{\partial S}{\partial z} = 0. \tag{2.23}$$

Recalling the definition of potential temperature $\theta = T(p/p_0)^{-\kappa}$, we have

$$S = c_p \log T - \Re \log p + \text{constant}$$

so that, using (2.22) and the hydrostatic equation, we get

$$\frac{\partial S}{\partial z} = \frac{g}{T}. \tag{2.24}$$

Now take the vertical derivative of (2.23) to obtain

$$\frac{\partial}{\partial t}\left(\frac{g}{T}\right) + \mathbf{V} \cdot \nabla \left(\frac{g}{T}\right) + \frac{\partial \mathbf{V}}{\partial z} \cdot \nabla \left(c_p \log T - \Re \log p\right) + \frac{g}{T}\frac{\partial w}{\partial z} = 0$$

which, by simple rearrangement of terms, may be expressed

$$\frac{\partial T}{\partial t} = \frac{c_p T}{g}\frac{\partial \mathbf{V}}{\partial z} \cdot \nabla T - \mathbf{V} \cdot \nabla T - \frac{T}{g\rho}\frac{\partial \mathbf{V}}{\partial z} \cdot \nabla p + T\frac{\partial w}{\partial z}. \tag{2.25}$$

At this point the geostrophic wind approximation and its vertical derivative, the thermal wind equation, are introduced:

$$\mathbf{V}_g = \frac{1}{f\rho}\mathbf{k} \times \nabla p; \qquad \frac{\partial \mathbf{V}_g}{\partial z} = \frac{g}{fT}\mathbf{k} \times \nabla T.$$

If these are substituted into (2.25), the first right-hand term vanishes and the second and third terms on that side cancel, leaving the simple relationship

$$\frac{\partial T}{\partial t} = T \frac{\partial w}{\partial z}. \tag{2.26}$$

This equation is sufficient for predicting the stratospheric temperature as long as the assumptions of geostrophy and vertical isothermy are acceptable. We will use this simple prognostic equation in the sequel.

2.5 Pressure co-ordinates

We have used geometric height as the vertical co-ordinate. This seems the obvious choice, and was the one made by Richardson. However, he also briefly considered the possibility of interchanging the roles of pressure and height and of using isobaric co-ordinates. If pressure is treated as an independent variable, the heights of the isobaric surfaces become dependent variables, and the rate of change of pressure following the flow replaces w as a measure of vertical velocity. Since ρ and θ are functions only of p and T, isotherms on an isobaric surface are also lines of constant density and potential temperature, facilitating analysis. In the barotropic case, where ρ is a function of pressure only, isobaric (constant p), isothermal (constant T), isopycnal (constant ρ) and isentropic (constant S) surfaces all coincide.

Isobaric co-ordinates were used by Vilhelm Bjerknes in preparing his synoptic charts. He plotted the heights and other variables on ten sheets at the standard levels from 100 hPa to 1000 hPa. Richardson discussed this (*WPNP*, p. 17) and wrote 'This system readily yields elegant approximations.' But he considered that deformable co-ordinate surfaces that vary in time would be inconvenient. He set down the expressions for the relationships between the derivatives in the two systems:

$$\left(\frac{\partial \psi}{\partial x}\right)_z = \left(\frac{\partial \psi}{\partial x}\right)_p - \frac{\partial \psi}{\partial z}\left(\frac{\partial z}{\partial x}\right)_p \tag{2.27}$$

$$\left(\frac{\partial \psi}{\partial y}\right)_z = \left(\frac{\partial \psi}{\partial y}\right)_p - \frac{\partial \psi}{\partial z}\left(\frac{\partial z}{\partial y}\right)_p. \tag{2.28}$$

He stated that 'The result of these substitutions is to produce a large number of terms. The additional terms are small, but they are not always negligible in comparison with the errors of observations. ... On this account I have preferred to use instead ... [height co-ordinates].' In fact, since the Lagrangian time-derivative is invariant with respect to the co-ordinate transformation, we can write

$$\frac{d\psi}{dt} = \left(\frac{\partial \psi}{\partial t}\right)_p + u\left(\frac{\partial \psi}{\partial x}\right)_p + v\left(\frac{\partial \psi}{\partial y}\right)_p + \omega\frac{\partial \psi}{\partial p},$$

where $\omega = dp/dt$. This is no more complicated than the corresponding expansion in z-co-ordinates. Thus, Richardson's concern with small additional terms appears to be unwarranted.[5]

Of greater interest is Richardson's allusion to 'elegant approximations'. What can he have meant? From (2.27) and (2.28) and the hydrostatic relation, it follows that the pressure gradient force in p-co-ordinates is simpler than in z-co-ordinates:

$$\frac{1}{\rho}\nabla_z p = g\nabla_p z.$$

The density does not appear on the right, so this term is linear in the dependent variables. The expression for the geostrophic wind is similarly simplified:

$$\mathbf{V}_g = \frac{g}{f}\mathbf{k} \times \nabla_p z$$

and the thermal wind has only a single term

$$\frac{\partial \mathbf{V}_g}{\partial z} = \frac{\Re_{gas}}{f}\mathbf{k} \times \nabla_p T$$

which, again, is simpler and more elegant than for height co-ordinates.

The continuity equation takes a particularly elegant form in pressure co-ordinates, becoming a diagnostic equation. We recall Equation (2.9), which can be written

$$\frac{\partial \rho}{\partial t} + \mathbf{V} \cdot \nabla_z \rho + \rho \nabla_z \cdot \mathbf{V} + \frac{\partial \rho w}{\partial z} = 0.$$

Using the hydrostatic equation to replace ρ by p we can write

$$\frac{\partial}{\partial p}\left\{\frac{\partial p}{\partial t} + \mathbf{V} \cdot \nabla_z p + w\frac{\partial p}{\partial z}\right\} + \left[\nabla_z \cdot \mathbf{V} - \frac{\partial \mathbf{V}}{\partial p} \cdot \nabla_z p\right] = 0.$$

The term in braces is just $\omega = dp/dt$. By (2.27) and (2.28) the term in square brackets is the horizontal divergence in isobaric co-ordinates. Thus we get the remarkable result:

$$\nabla_p \cdot \mathbf{V} + \frac{\partial \omega}{\partial p} = 0. \tag{2.29}$$

The continuity equation contains no explicit time derivative; it is a diagnostic equation. It is formally identical to the continuity equation for an incompressible fluid. The tendency equation in pressure co-ordinates is also greatly simplified. Integrating (2.29) and using the boundary condition $\omega = 0$ at $p = 0$, we get

$$\omega = -\int_0^p \nabla_p \cdot \mathbf{V}\,dp, \tag{2.30}$$

[5] Eliassen (1949) showed (in his §§20, 21) that there are additional terms in the continuity equation ((2.29) below), but that they may be disregarded in general.

which is notably simpler than (2.15). It shows that ω is determined completely once **V** is known.

An intriguing question is whether Richardson could possibly have discovered the isobaric form of the continuity equation. Platzman (1967) argues that if he had discovered it he would have presented it in his book and would indeed have used it. It is difficult to dispute this view, particularly when Sydney Chapman's words about Richardson are recalled: 'He once told me that he had put all he knew into the book' (*WPNP*, Dover Edition, p. viii). The equation was derived by Eliassen in his seminal paper of 1949 on the quasi-geostrophic aproximation. It had appeared earlier, in work of Sutcliffe and of a Belgian meteorologist, O. Godard. However, more recently, Eliassen (1999) pointed out what had not been previously realised: the diagnostic relationship in pressure co-ordinates was discovered by Vilhelm Bjerknes, and is to be found, not as a mathematical equation but in textual form, in Bjerknes' *Kinematics* (1911). As Richardson was very familiar with Bjerknes' work, it is indeed conceivable that he knew about the equation and that this was what he meant when he wrote that the pressure system 'yields elegant approximations'. But until we find more conclusive documentary evidence, the matter remains unclear.

Another co-ordinate system introduced by Eliassen (1949) was the log-pressure system. If we define $Z = -H \log(p/p_0)$, where H and p_0 are constants, Z is a re-labelling of the pressure surfaces. With H chosen as the scale-height of the atmosphere and p_0 as the standard surface pressure, Z is approximately equal to the geometric height z. This system was used to great effect by Holton (1975) in his monograph on the stratosphere and mesosphere.

The lower boundary condition in pressure co-ordinates is more complicated than in height co-ordinates, because the pressure at the Earth's surface changes with time. This property offsets to some extent the advantages of the pressure system. Phillips (1953) introduced normalised pressure $\sigma = p/p_s$ as the so-called sigma-co-ordinate. This yields the particularly simple lower boundary condition $\dot{\sigma} = 0$ at $\sigma = 1$. It has been used extensively in numerical modelling of the atmosphere. Richardson hinted at the possibility of terrain-following co-ordinates (*WPNP*, p. 92) but did not follow up on this idea.

3

The oscillations of the atmosphere

> The most interesting result is the existence of a second class of free
> oscillations besides those whose existence may be at once inferred by
> analogy from the simpler problem of the oscillations of an ocean covering
> a non-rotating globe. *(Hough, 1898)*

To gain a complete understanding of the causes of failure of Richardson's forecast, we need to study the rich variety of oscillations that the atmosphere can sustain. The most prominent regular motions are those associated with the diurnal variation in solar radiation, the thermal tides. Such oscillations have been studied for over two centuries; Cartwright (1999) provides an interesting historical account. Forced motions such as tides are responses to external influences and their frequencies are determined by the variation in time of the forcing. Of greater interest here are the free modes or 'natural oscillations' of the atmosphere: like many other physical systems, the air has a propensity to oscillate at certain preferred frequencies and in this chapter we will study these normal modes. We will confine our attention to an atmosphere in hydrostatic balance but note that recently Kasahara and Qian (2000) have studied the normal modes when this assumption is relaxed.

A relatively simple, self-contained system of equations can be obtained if we assume that the amplitude of the motion is so small that all nonlinear terms can be neglected. The horizontal structure is then governed by a system equivalent to the linear shallow water equations that describe the small-amplitude motions of a shallow layer of incompressible fluid. These equations were first derived by Laplace in his discussion of tides in the atmosphere and ocean, and are called the Laplace tidal equations. The simplest means of deriving the linear shallow water equations from the primitive equations is to assume that the vertical velocity vanishes identically. This is done in §3.1 below. However, this assumption precludes the possibility of studying internal modes for which $w \neq 0$. A more general derivation, based on a separation of the horizontal and vertical dependencies of the variables, is presented in Appendix 3.

3.1 The Laplace tidal equations

Let us assume that the motions under consideration can be described as small perturbations about a state of rest, in which the temperature is a constant, T_0, and the pressure $\bar{p}(z)$ and density $\bar{\rho}(z)$ vary only with height. The basic state variables satisfy the gas law and are in hydrostatic balance:

$$\bar{p} = \Re\bar{\rho}T_0, \qquad \frac{d\bar{p}}{dz} = -g\bar{\rho}.$$

The variations of mean pressure and density follow immediately:

$$\bar{p}(z) = p_0 \exp(-z/H), \qquad \bar{\rho}(z) = \rho_0 \exp(-z/H),$$

where p_0 and ρ_0 are the mean surface pressure and density and $H = p_0/g\rho_0 = \Re T_0/g$ is the scale-height of the atmosphere. We consider only motions for which the vertical component of velocity vanishes identically: $w \equiv 0$. Let u, v, p and ρ denote variations about the basic state, each of these being a small quantity. We assume the Earth's surface is perfectly flat, consistent with zero vertical velocity. If all the quadratic terms are omitted, the horizontal momentum, continuity and thermodynamic equations become

$$\frac{\partial \bar{\rho}u}{\partial t} - f\bar{\rho}v + \frac{\partial p}{\partial x} = 0 \tag{3.1}$$

$$\frac{\partial \bar{\rho}v}{\partial t} + f\bar{\rho}u + \frac{\partial p}{\partial y} = 0 \tag{3.2}$$

$$\frac{\partial \rho}{\partial t} + \nabla \cdot \bar{\rho}\mathbf{V} = 0 \tag{3.3}$$

$$\frac{1}{\gamma\bar{p}}\frac{\partial p}{\partial t} - \frac{1}{\bar{\rho}}\frac{\partial \rho}{\partial t} = 0 \tag{3.4}$$

Density can be eliminated from the continuity equation (3.3) by means of the thermodynamic equation (3.4). Now let us assume that the horizontal and vertical dependencies of the perturbation quantities are separable:

$$\left\{ \begin{matrix} \bar{\rho}u \\ \bar{\rho}v \\ p \end{matrix} \right\} = \left\{ \begin{matrix} U(x, y, t) \\ V(x, y, t) \\ P(x, y, t) \end{matrix} \right\} Z(z). \tag{3.5}$$

The momentum and continuity equations can then be written

$$\frac{\partial U}{\partial t} - fV + \frac{\partial P}{\partial x} = 0 \tag{3.6}$$

$$\frac{\partial V}{\partial t} + fU + \frac{\partial P}{\partial y} = 0 \tag{3.7}$$

$$\frac{\partial P}{\partial t} + (\gamma g H)\nabla \cdot \mathbf{V} = 0 \tag{3.8}$$

(here $\mathbf{V} = (U, V)$ is the momentum vector). This is a set of three equations for the three dependent variables U, V, and P. They are mathematically isomorphic to the Laplace tidal equations for an ocean of mean depth $h = \gamma H = \gamma \Re T_0/g$. The quantity h is therefore called the equivalent depth. There is no dependence in this system on the vertical co-ordinate z.

The vertical structure follows from the hydrostatic equation, together with the relationship $p = (\gamma g H)\rho$ implied by the thermodynamic equation. It is determined by

$$\frac{dZ}{dz} + \frac{Z}{\gamma H} = 0, \tag{3.9}$$

the solution of which is $Z = Z_0 \exp(-z/\gamma H)$, where Z_0 is the amplitude at $z = 0$. If we set $Z_0 = 1$, then U, V and P give the momentum and pressure fields at the Earth's surface. These variables all decay exponentially with height. It follows from (3.5) that u and v actually increase with height as $\exp(\kappa z/H)$, where $\kappa = 1 - 1/\gamma = \Re/c_p$, but the kinetic energy decays. Solutions with more general vertical structures, and with non-vanishing vertical velocity, are discussed in Appendix 3.

3.2 Normal modes of the atmosphere

We investigate some simple wave-like solutions of the Laplace tidal equations. Holton (1975) gives a more extensive analysis, including treatments of the equatorial and mid-latitude β-plane approximations. The equations are

$$\frac{\partial U}{\partial t} - fV + \frac{\partial P}{\partial x} = 0 \tag{3.10}$$

$$\frac{\partial V}{\partial t} + fU + \frac{\partial P}{\partial y} = 0 \tag{3.11}$$

$$\frac{\partial P}{\partial t} + gh\nabla \cdot \mathbf{V} = 0 \tag{3.12}$$

where h is the mean fluid depth. These are the equations used by Richardson for his 'introductory example' (*WPNP*, Chapter 2). He showed that they can be reduced to a single equation for the pressure variable P. We can use the momentum equations to express U and V in terms of P:

$$\mathcal{D}^2 U \equiv \left(\frac{\partial^2}{\partial t^2} + f^2\right) U = -(P_{xt} + f P_y)$$

$$\mathcal{D}^2 V \equiv \left(\frac{\partial^2}{\partial t^2} + f^2\right) V = +(f P_x - P_{yt})$$

We now apply the operator $\mathcal{D}^2$ to the continuity equation and use these expressions to eliminate $\mathcal{D}^2 U$ and $\mathcal{D}^2 V$. The result is

$$\mathcal{D}^2 P_t + gh\{-(\nabla^2 P)_t + \beta P_x - 2\beta f V\}$$

where $\beta = df/dy = 2\Omega \cos\phi/a$. The remaining term with V arises because $\mathcal{D}^2$ does not commute with ∇^2. To eliminate it, we must apply $\mathcal{D}^2$ again; this introduces additional solutions, which may be ignored. The final equation is

$$\left(\frac{\partial^2}{\partial t^2} + f^2\right)^2 \frac{\partial P}{\partial t} + gh\left\{\left(\frac{\partial^2}{\partial t^2} + f^2\right)\left[-\frac{\partial}{\partial t}\nabla^2 P + \beta\frac{\partial P}{\partial x}\right] + 2\beta f\left(\frac{\partial^2 P}{\partial y \partial t} - f\frac{\partial P}{\partial x}\right)\right\} = 0.$$

$$(3.13)$$

This is the equation on page 15 of *WPNP*. We have not seen it written in this form elsewhere (normally, the longitudinal and time dependence are assumed to be harmonic, reducing it to an ordinary differential equation). Richardson comments wistfully on the equation: 'No wonder that isobaric maps look complicated, if this be their differential equation, derived from most stringently restricted hypotheses.' It is a fifth-order equation in the time derivative but, as we have said, additional solutions have been introduced by the application of the operator $\mathcal{D}^2$. We note that the Laplace tidal equations equations involve only three time derivatives.[1]

Vorticity and divergence

Before discussing the general solutions of (3.10)–(3.12), we will examine some enlightening limiting cases. By means of the Helmholtz Theorem, a general horizontal wind field $\mathbf{V}$ may be partitioned into rotational and divergent components

$$\mathbf{V} = \mathbf{V}_\psi + \mathbf{V}_\chi = \mathbf{k} \times \nabla\psi + \nabla\chi.$$

If the vorticity ζ and divergence δ are defined by

$$\zeta = \mathbf{k} \cdot \nabla \times \mathbf{V}, \qquad \delta = \nabla \cdot \mathbf{V},$$

---

[1] As pointed out to the author by Akira Kasahara, if P and U are eliminated, the resulting equation for V is third order in time.

then the stream function ψ and velocity potential χ are related to the vorticity and divergence by the Poisson equations

$$\nabla^2 \psi = \zeta, \qquad \nabla^2 \chi = \delta.$$

From the momentum equations it is straightforward to derive equations for the vorticity and divergence tendencies. Together with the continuity equation, they are

$$\frac{\partial \zeta}{\partial t} + f\delta + \beta v = 0 \tag{3.14}$$

$$\frac{\partial \delta}{\partial t} - f\zeta + \beta u + \nabla^2 P = 0 \tag{3.15}$$

$$\frac{\partial P}{\partial t} + gh\delta = 0. \tag{3.16}$$

These equations are completely equivalent to (3.10)–(3.12); no additional approximations have yet been made. However, the vorticity and divergence forms enable us to examine various simple approximate solutions.

Solutions when rotation vanishes

We consider solutions in the case of no rotation, $\Omega = f = \beta = 0$. The pressure field P must then satisfy the classical wave equation

$$\frac{\partial^2 P}{\partial t^2} - gh\nabla^2 P = 0. \tag{3.17}$$

The solutions of this equation can be found using the well-known eigenvalues and eigenfunctions of the Laplacian operator, given by

$$\nabla^2 Y_n^m(\lambda, \phi) = -\frac{n(n+1)}{a^2} Y_n^m(\lambda, \phi)$$

where $Y_n^m(\lambda, \phi) = e^{im\lambda} P_n^m(\sin \phi)$ is a spherical harmonic function. The total wavenumber n determines the spatial scale: the larger n is, the smaller is the horizontal scale. Assuming the pressure P has the wave-like structure of a spherical harmonic, we substitute the expression $P = P_0 Y_n^m(\lambda, \phi) \exp(-i\nu t)$ in (3.17) and immediately deduce an expression for the frequency:

$$\nu = \pm\nu_G \quad \text{where} \quad \nu_G \equiv \sqrt{\frac{n(n+1)gh}{a^2}}, \tag{3.18}$$

the classical frequency of *pure gravity waves*. These waves can travel either eastward or westward, at high speed. In fact, with no rotation there is no preferred axis so these waves may travel in any direction. Taking typical values $g = 10\,\text{m s}^{-2}$, $h = 10\,\text{km}$ and $a = 6370\,\text{km}$, the lowest frequency (that for $n = 1$) is $7.02 \times 10^{-5}\,\text{s}^{-1}$, which is quite close to the frequency $\Omega = 7.27 \times 10^{-5}\,\text{s}^{-1}$ associated with the Earth's rotation. The frequency increases roughly in proportion to n. The vorticity vanishes identically for these gravity wave solutions.

We now seek steady-state solutions $(\partial/\partial t \equiv 0)$ in the case of no rotation. The equations then require P to be independent of both x and y, and therefore constant, and the wind to be non-divergent. An arbitrary unchanging velocity field $\mathbf{V}$ satisfying $\nabla \cdot \mathbf{V} = 0$ is a solution of (3.14)–(3.16) for an undisturbed pressure field in the case of no rotation.

We have found two types of solution in the extreme case of vanishing rotation, one without vorticity, one without divergence. We will consider how these are related to more general solutions. Lamb (1932, §223) makes the following observations:

To understand the nature of the *free* oscillations, it is best to begin with the case of no rotation ($\Omega = 0$). As Ω is increased, the pairs of numerically equal, but oppositely signed, values of [frequency] ... begin to diverge in absolute value, that being the greater which has the same sign with Ω. The character of the fundamental modes is also gradually altered. These oscillations are distinguished as 'of the First Class'.

At the same time certain steady motions which are possible, without change of level, when there is no rotation, are converted into long-period oscillations with change of level, the ... [frequencies] being initially comparable with Ω. The corresponding modes are designated as 'of the Second Class'.

We consider next the two special cases in which either the divergence or the vorticity is very small, and see how Lamb's remarks are to be interpreted.

Rossby–Haurwitz modes

We suppose that the solution is quasi-nondivergent, i.e., we assume $|\delta| \ll |\zeta|$, so the wind is given approximately in terms of the stream function $(u, v) \approx (-\psi_y, \psi_x)$. The vorticity equation becomes

$$\nabla^2 \psi_t + \beta \psi_x = O(\delta), \tag{3.19}$$

and we can ignore the right-hand side. Assuming the stream function has the wave-like structure of a spherical harmonic, we substitute the expression $\psi = \psi_0 Y_n^m(\lambda, \phi)\exp(-i\nu t)$ in the vorticity equation and immediately deduce an expression for the frequency:

$$\nu = \nu_R \equiv -\frac{2\Omega m}{n(n+1)}. \tag{3.20}$$

This is the celebrated dispersion relation for *Rossby–Haurwitz waves* (Haurwitz, 1940). If we ignore sphericity (the β-plane approximation) and assume harmonic dependence $\psi(x, y, t) = \psi_0 \exp[i(kx + \ell y - \nu t)]$, then (3.19) has the dispersion relation

$$c = \frac{\nu}{k} = -\frac{\beta}{k^2 + \ell^2},$$

which is the expression for phase-speed found by Rossby (1939).[2] The Rossby or Rossby–Haurwitz waves are, to the first approximation, non-divergent waves that travel westward, the phase speed being greatest for the waves of largest scale. They are of relatively low frequency, and the frequency decreases as the spatial scale decreases. We note that (3.20) implies that $|\nu| \leq \Omega$. In the limit $\Omega \to 0$, the frequency vanishes and we obtain the steady-state non-divergent solutions found above.

To the same degree of approximation, we may write the divergence equation (3.15) as

$$\nabla^2 P - f\zeta - \beta\psi_y = O(\delta). \tag{3.21}$$

Ignoring the right-hand side, we get the linear balance equation

$$\nabla^2 P = \nabla \cdot f\nabla\psi, \tag{3.22}$$

a diagnostic relationship between the geopotential and the stream function. This relationship also follows immediately from the assumption that the wind is both non-divergent ($\mathbf{V} = \mathbf{k} \times \nabla\psi$) and geostrophic ($f\mathbf{V} = \mathbf{k} \times \nabla P$). If variations of f are ignored, we can assume $P = f\psi$. The wind and pressure are in approximate geostrophic balance for Rossby–Haurwitz waves.

Gravity wave modes

If we assume now that the solution is quasi-irrotational, i.e. that $|\zeta| \ll |\delta|$, then the wind is given approximately by $(u, v) \approx (\chi_x, \chi_y)$ and the divergence equation becomes

$$\nabla^2 \chi_t + \beta\chi_x + \nabla^2 P = O(\zeta)$$

with the right-hand side negligible. Using the continuity equation to eliminate P, we get

$$\nabla^2 \chi_{tt} + \beta\chi_{xt} - gh\nabla^4 \chi = 0.$$

Seeking a solution $\chi = \chi_0 Y_n^m(\lambda, \phi)\exp(-i\nu t)$, we find that

$$\nu^2 + \left(-\frac{2\Omega m}{n(n+1)}\right)\nu - \frac{n(n+1)gh}{a^2} = 0. \tag{3.23}$$

---

[2] Actually, Rossby considered only the case $\ell = 0$ of waves with infinite lateral extent. He also allowed for a mean zonal velocity, $\bar{u}$, which Doppler-shifts the wave solution.

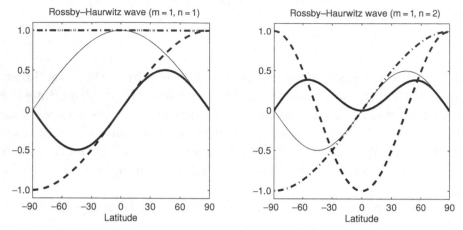

Figure 3.1 Latitudinal structure of the first two Rossby–Haurwitz waves for $m = 1$. Stream function (thin line), zonal wind u (dashed line), meridional wind v (dot-dashed line) and pressure (thick line). The first mode (left panel, $m = 1, n = 1$) has a period of precisely one day. The second mode (right panel, $m = 1, n = 2$) has period of three days.

The coefficient of the second term is just the Rossby–Haurwitz frequency ν_R found in (3.20) above, so that

$$\nu = \pm\sqrt{\nu_G^2 + \left(\tfrac{1}{2}\nu_R\right)^2} - \tfrac{1}{2}\nu_R.$$

Noting that $|\nu_G| \gg |\nu_R|$, we can write the two roots

$$\nu_+ \approx \nu_G - \tfrac{1}{2}\nu_R, \qquad \nu_- \approx -\nu_G - \tfrac{1}{2}\nu_R,$$

so that $|\nu_+|$ is slightly greater than $|\nu_-|$, as indicated by Lamb. If we ignore ν_R, it follows that

$$\nu_\pm = \pm\nu_G \equiv \pm\sqrt{\frac{n(n + 1)gh}{a^2}},$$

the frequency of pure gravity waves. There are then two solutions, representing waves travelling eastward and westward with equal speeds. The frequency increases approximately linearly with the total wavenumber n.

Two particular solutions

It is useful to examine a few special cases. The latitudinal structures of the first two Rossby–Haurwitz waves for $m = 1$ are plotted in Fig. 3.1. The stream function (thin line) is proportional to $P_n^m(\sin\phi)$ and the winds u (dashed line) and v (dot-dashed line) are derived directly from this. The wind is non-divergent. The pressure (thick line) is derived assuming $p = f\psi$. The first mode (left panel, $m = 1, n = 1$) has a period of precisely one day (this follows from (3.20)). It retrogresses (moves westward) at the same rate that the Earth is spinning eastward and is therefore stationary in an absolute frame. The pressure is antisymmetric about the equator. This mode is a limiting case of what is called the *mixed Rossby-gravity wave*.

The second mode (right panel, Fig. 3.1) has opposite parity: the pressure field is now symmetric about the equator. Since $m = 1$ and $n = 2$, the frequency (by (3.20)) is $\Omega/3$. We might call it a three-day wave. However, in the atmosphere the corresponding mode is modified significantly by divergence and, to a lesser extent, doppler-shifted by the predominantly westerly flow. The frequency is reduced and the period extended, giving the mode the title 'five-day wave'. This mode is of central relevance to Richardson's barotropic forecast, considered in the following chapter.

3.3 Atmospheric tides

Ocean tides have been known for thousands of years. The discovery of tides in the atmosphere had to await the development of the barometer. While the ocean tides are forced by gravity, the air tides are due primarily to the thermal forcing that results from the diurnal variations in sunshine and are therefore synchronised with local solar time, not lunar time. The great German naturalist and traveller Alexander von Humboldt took a barometer on his travels to South America (c. 1800) and noticed that the semi-diurnal oscillations of pressure were so regular in the tropics that he could regard the barometer as a sort of clock (Chapman and Lindzen, 1970). Humboldt referred to the earlier observations of Robert de Lamanon who, in 1785, observed the pressure hourly for a continuous period of 75 hours while crossing the equator and noticed the twice-daily variations with maxima at about 10 a.m. and 10 p.m. local time. Lamanon sent an account back to the *Académie Royale* in Paris, but he himself never returned, being one of a party of men massacred shortly afterwards by the natives of Samoa (Cartwright, 1999).

Laplace showed that for an isothermal atmosphere of scale height H the tidal oscillations could be inferred from those of an incompressible ocean of depth H. Significant progress in analysing the tidal equations was made by Margules and Hough. Margules (1893) identified two groups of solutions with distinct characteristics. Slightly later, and quite independently, Hough (1898) carried out a closely related study in which he expanded the solutions in series of spherical harmonics and again determined two distinct types of solution: low frequency rotational modes for which vorticity dominates divergence and high frequency gravity modes with divergence dominating. In recognition of this work, the eigensolutions of the Laplace tidal equations are called Hough functions.

Two extensive numerical analyses of the Laplace tidal equations were carried out in the 1960s. Flattery (1967), using an expansion similar to that of Hough, expressed the coefficients as convergent continued fractions (the recursion relation is of third order; for the usual power series expansion it is of fifth order, which defies analysis). He tabulated the solutions of most importance in tidal theory. He also discussed three simple special cases in which solutions are known in closed form. Longuet-Higgins (1968) formulated a matrix eigenvalue problem, which he solved numerically. He depicted a large number of solutions in graphical and tabular form and discussed several asymptotic limiting cases. Dikii (1965) also studied the atmospheric oscillation problem both numerically and analytically. In a series of papers starting with Kasahara (1976), the solutions of most relevance to free atmospheric oscillations were discussed. We will present Kasahara's solution method below.

3.4 Numerical solution of the Laplace tidal equations

The tidal equations have resisted a complete analysis up to now, due to their subtle mathematical nature. Their eigenstructure has some unusual characteristics, which are of mathematical as well as geophysical interest. We consider in this section the mathematical difficulties of solving the equation (3.13), present Kasahara's (1976) numerical method and discuss the asymptotic behaviour of the solutions.

Since the coefficients are independent of the time and longitude, we can assume that the solution has the form

$$P(\lambda, \phi, t) = Y(\phi) \exp[i(m\lambda - 2\Omega\sigma t)]$$

and substitute this into (3.13) to obtain a differential equation for the meridional structure $Y(\phi)$. After a little algebra we get

$$\frac{1}{\cos\phi} \frac{\partial}{\partial\phi} \left[\frac{\cos\phi}{(\sigma^2 - \sin^2\phi)} \frac{\partial Y}{\partial\phi} \right] + \left\{ \frac{1}{\sigma^2 - \sin^2\phi} \left[\frac{m}{\sigma} \frac{\sigma^2 + \sin^2\phi}{\sigma^2 - \sin^2\phi} - \frac{m^2}{\cos^2\phi} \right] + \epsilon \right\} Y = 0,$$

where $\epsilon = (2\Omega a)^2/gh$ is known as Lamb's parameter. If we define $\mu = \sin\phi$, the equation may be written

$$\frac{d}{d\mu} \left[\frac{1 - \mu^2}{\sigma^2 - \mu^2} \frac{dY}{d\mu} \right] + \left\{ \frac{1}{\sigma^2 - \mu^2} \left[\frac{m}{\sigma} \frac{\sigma^2 + \mu^2}{\sigma^2 - \mu^2} - \frac{m^2}{1 - \mu^2} \right] + \epsilon \right\} Y = 0. \quad (3.24)$$

We have written total derivatives since $Y = Y(\mu)$. This is the form found in Holton, (1975, p. 61). The meridional structure of the normal modes is determined by the eigensolutions of this second order o.d.e., to which we will refer briefly as the Laplace tidal equation (LTE). Boundary conditions require Y to be regular at the geographic poles. Two types of problem involving the LTE arise in geophysics. For forced oscillations, the frequency σ of the forcing is specified and we look for a response of similar frequency. The eigenvalue is ϵ. For each ϵ_n we deduce an 'equivalent depth' h_n, and calculate the corresponding vertical structure. We remark that h_n can be positive or negative. For free oscillations, the equivalent depth h is an eigenvalue of the homogeneous vertical structure equation. The LTE is then solved, with ϵ determined from h, for the frequency σ as the eigenvalue and Y as the eigenfunction.

3.4.1 Mathematical difficulties

The standard form of the Sturm–Liouville equation is

$$\frac{d}{d\mu}\left(p(\mu)\frac{dY}{d\mu}\right) + [q(\mu) + \lambda r(\mu)]Y = 0 \qquad (3.25)$$

where $p(\mu)$ is regular and has *no zeros within the domain* (it may vanish at the boundaries). We may assume that both $p(\mu)$ and $r(\mu)$ are positive. The properties of the solutions of this equation for various boundary conditions are derived in Morse and Feshbach (1953, pp. 719, *et seq.*):

(i) The equation is self-adjoint and the eigenvalues λ are real.
(ii) The eigenfunctions for different values of λ are orthogonal.
(iii) The eigenfunctions form a complete set.
(iv) There is a denumerable infinity of non-negative eigenvalues with a single limit point at $+\infty$.
(v) The zeros of the eigenfunctions behave according to the Sturmian oscillation theorems.

For the LTE, $p(\mu) = (1 - \mu^2)/(\sigma^2 - \mu^2)$ blows up at the 'critical latutudes' where $\mu = \pm\sigma$, and the equation is singular for $|\sigma| < 1$. Normally this would imply singularity of the solution there. But in fact all solutions are regular throughout the domain. The critical latitudes were shown by Brillouin (1932) to be apparent singularities. However, although the solutions are regular at the critical points, their existence greatly complicates the numerical and asymptotic analysis of the LTE.

Since the LTE cannot be written in standard Sturm–Liouville form, there is no guarantee that the five properties listed above hold good. It has been shown that the eigenvalues ϵ_n of the LTE are real and the eigenfunctions for distinct ϵ are orthogonal. It has also been proved by Holl (1970) that they form a complete set. However, the fourth and fifth properties do not hold. For $|\sigma| < 1$ there is a double infinity of eigenvalues, with limit points at both $+\infty$ and $-\infty$. Furthermore, the zeros for successive eigenfunctions do not behave in quite as simple a manner as those for a regular Sturm–Liouville problem. This question is discussed by Longuet-Higgins (1968).

3.4.2 Kasahara's solution method

The mathematical complications associated with the apparent singularities make a direct assault on the Laplace tidal equations ineffectual, and more circuitous methods have been devised. Here we outline the approach of Kasahara (1976). Since we are interested in free oscillations, we will specify the equivalent depth and solve the LTE for frequency. We introduce a stream function ψ and velocity potential χ as in §3.2 above, and assume the dependent variables have the form

$$\begin{Bmatrix} \chi \\ \psi \\ P \end{Bmatrix} = \begin{Bmatrix} (2\Omega a^2)\hat{\chi} \\ (2\Omega a^2)\hat{\psi} \\ (2\Omega a)^2 \hat{P} \end{Bmatrix} \exp[i(m\lambda - 2\Omega\sigma t)]. \tag{3.26}$$

The system of three equations (3.14)–(3.16) now becomes

$$(\sigma\nabla^2 - m)i\,\hat{\chi} + (\mu\nabla^2 + D)\hat{\psi} - \nabla^2\hat{P} = 0 \tag{3.27}$$

$$(\sigma\nabla^2 - m)\hat{\psi} + (\mu\nabla^2 + D)i\,\hat{\chi} = 0 \tag{3.28}$$

$$\nabla^2 i\,\hat{\chi} + \epsilon\sigma\hat{P} = 0 \tag{3.29}$$

where the differential operators are defined as

$$D = (1 - \mu^2)\frac{d}{d\mu}, \qquad \nabla^2 = \frac{d}{d\mu}\left[(1 - \mu^2)\frac{d}{d\mu}\right] - \frac{m^2}{1 - \mu^2}.$$

To solve this system for a given m, we expand in Legendre functions:

$$\begin{Bmatrix} \hat{\chi} \\ \hat{\psi} \\ \hat{P} \end{Bmatrix} = \sum_{n=m}^{N} \begin{Bmatrix} i A_n^m \\ B_n^m \\ C_n^m \end{Bmatrix} P_n^m(\mu). \tag{3.30}$$

We substitute this and equate coefficients of P_n^m to zero. The system separates into two subsystems, corresponding to symmetric and anti-symmetric solutions. The symmetric case is expressible as

$$(\mathbf{A} - \sigma\mathbf{I})\mathbf{X} = \mathbf{0}, \tag{3.31}$$

where the vector $\mathbf{X}$ of coefficients is defined as

$$\mathbf{X}^T = \left(A_m^m, B_{m+1}^m, C_m^m, \ldots, A_{m+2N}^m, B_{m+2N+1}^m, C_{m+2N}^m \right)$$

and the pentadiagonal matrix $\mathbf{A}$ is given by:

$$\mathbf{A} = \begin{pmatrix}
K_m & P_{m+1} & -1 & 0 & 0 & 0 & \cdots \\
q_m & K_{m+1} & 0 & P_{m+2} & 0 & 0 & \cdots \\
r_m & 0 & 0 & 0 & 0 & 0 & \cdots \\
0 & q_{m+1} & 0 & K_{m+2} & P_{m+3} & -1 & \cdots \\
0 & 0 & 0 & q_{m+2} & K_{m+3} & 0 & \cdots \\
0 & 0 & 0 & r_{m+2} & 0 & 0 & \cdots \\
\vdots & \vdots & \vdots & \vdots & \vdots & \vdots & \ddots
\end{pmatrix}$$

where

$$K_n = \frac{-m}{n(n+1)}, \quad p_n = \frac{(n+1)(n+m)}{n(2n+1)}, \quad q_n = \frac{n(n-m+1)}{(n+1)(2n+1)}, \quad r_n = \frac{-n(n+1)}{\epsilon}.$$

The anti-symmetric case is expressible as

$$(\mathsf{B} - \sigma \mathsf{I})\mathbf{Y} = 0, \tag{3.32}$$

where the vector $\mathbf{Y}$ of coefficients is defined as

$$\mathbf{Y}^T = \left(B_m^m, A_{m+1}^m, C_{m+1}^m, \ldots, B_{m+2N}^m, A_{m+2N+1}^m, C_{m+2N+1}^m \right)$$

and the pentadiagonal matrix B is given by:

$$\mathsf{B} = \begin{pmatrix} K_m & p_{m+1} & 0 & 0 & 0 & 0 & \cdots \\ q_m & K_{m+1} & -1 & p_{m+2} & 0 & 0 & \cdots \\ 0 & r_{m+1} & 0 & 0 & 0 & 0 & \cdots \\ 0 & q_{m+1} & 0 & K_{m+2} & p_{m+3} & 0 & \cdots \\ 0 & 0 & 0 & q_{m+2} & K_{m+3} & -1 & \cdots \\ 0 & 0 & 0 & 0 & r_{m+3} & 0 & \cdots \\ \vdots & \vdots & \vdots & \vdots & \vdots & \vdots & \ddots \end{pmatrix}.$$

The dimensionless frequency σ is determined as the eigenvalue of matrix A or B and the corresponding eigenvector $\mathbf{X}$ or $\mathbf{Y}$ determines the coefficients of the series (3.30). Fuller details may be found in Kasahara (1976) together with FORTRAN code to generate the solutions.[3] The algebraic eigenvalue equations (3.31) and (3.32) are in standard form and may immediately be solved using one of the widely available eigensystem packages. We mention also Swarztrauber and Kasahara (1985), which describes computer code to calculate the frequency and structure functions for positive values of the equivalent depth.

3.4.3 Asymptotic behaviour and the nondivergent limit

An extensive asymptotic analysis of the LTE was presented in Longuet-Higgins (1968). We will just make a few observations here. The eigenvalues of the LTE for zonal wavenumber one, as functions of the square root of Lamb's parameter $\sqrt{\epsilon}$, are shown in Fig. 3.2. The complexity of the figure indicates the richness of the eigenstructure of the LTE. The westward travelling waves of the second class are the only ones with finite frequency σ as $\epsilon \to 0$. Recall $\nu = 2\Omega\sigma$, so the dimensional frequency tends to zero. This limit represents nondivergent flow

---

[3] Note that there is a FORTRAN statement T3=T1+2 missing, immediately following the statement T2=T1+1 near the end of the subroutine AMTRX (Kasahara, 1976, p. 689).

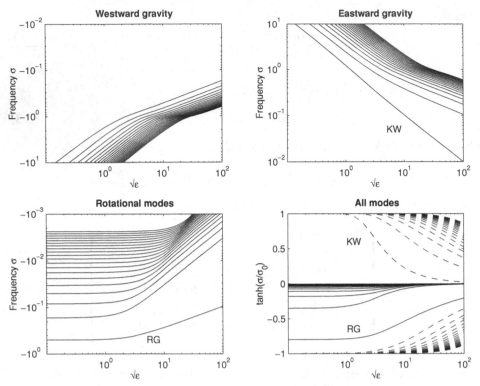

Figure 3.2 Eigenfrequencies σ of the Laplace tidal equations as functions of $\sqrt{\epsilon}$ for $m = 1$. For westward gravity (upper left), eastward gravity (upper right) and rotational modes (lower left) a log-log scale is used. The lower right panel shows all the modes; the ordinate is $\tanh(\sigma/\sigma_0)$, where $\sigma_0 = \frac{1}{4}\tanh^{-1}\frac{1}{2}$.

and the solutions are the Rossby–Haurwitz waves. As ϵ increases the first of these changes its asymptotic behaviour and, in the limit of large ϵ behaves like a solution of the first class; it is the mixed Rossby-gravity wave, marked RG in the figure. In a somewhat analogous manner, the behaviour of the first eastward travelling gravity wave changes as ϵ increases; this mode is called the Kelvin wave (marked KW). The Rossby-gravity and Kelvin waves are important in atmospheric dynamics, and they have been used in mechanistic models of the quasi-biennial oscillation (QBO) of the stratosphere.

Nondivergent flow corresponds to the limit $h \to \infty$ or $\epsilon \to 0$. Since $\hat{\chi}$ vanishes, the vorticity equation reduces to

$$(\sigma \nabla^2 - m)\hat{\psi} = 0,$$

which is just the associated Legendre equation, with solution

$$\hat{\psi} = P_n^m(\mu), \qquad \sigma = -\frac{m}{n(n+1)}.$$

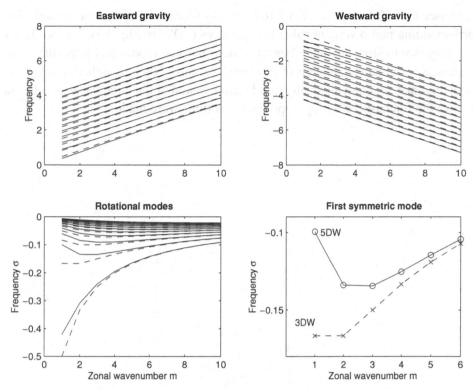

Figure 3.3 Eigenfrequencies of the LTE as functions of zonal wavenumber m. Solid lines are the calculated frequencies with $\epsilon = 8.7484$ and dashed lines are the approximate values based on (3.18) and (3.20). Eastward and westward gravity waves are shown in the top panels, and rotational modes in the lower left panel. The lower right panel is for the first symmetric rotational mode.

This is the Rossby–Haurwitz wave obtained already by simple approximation. The high-frequency solutions are excluded by the assumption of non-divergence.

An isothermal atmosphere with T = 244 K has an equivalent depth $h = 10$ km, and in this case $\epsilon = 8.7484$. The free oscillations or normal modes are obtained by fixing ϵ and solving the LTE with σ as the eigenvalue. This case was studied by Kasahara (1976). In Fig. 3.3 we show the frequencies calculated by solving (3.31) and its anti-symmetric counterpart (3.32) for a range of wavenumbers. The solid lines are the calculated frequencies of the Hough modes for $h = 10$ km. The dashed lines are the approximate values based on (3.18) for the gravity waves and (3.20) for the nondivergent Rossby–Haurwitz waves. For the gravity waves (top panels) the approximate values are quite accurate except for the lowest wavenumbers or slowest modes. For the rotational modes (lower left panel) the disagreement is more pronounced, especially for zonal wavenumber one. The lower right panel zooms in on the values for the first symmetric rotational mode. This Hough mode has frequency $\sigma \approx 0.1$ or period close to five days (denoted 5DW in the figure). The corresponding non-divergent value is three days (3DW in Fig. 3.3). The tendency of divergence to slow down the largest-scale rotational modes had important consequences for practical weather forecasting: the early NWP models did not allow for divergence, and unrealistically rapid retrogression of the long waves was found to spoil the forecasts unless appropriate corrections were made.

4

The barotropic forecast

> Before attending to the complexities of the actual atmosphere . . . it may
> be well to exhibit the working of a much simplified case.
>
> (*WPNP*, p. 4)

To clarify the essential steps required to obtain a numerical solution of the equations
of motion, Richardson included in his book an introductory example in which he
integrated a system equivalent to the linearised shallow water equations (*WPNP*,
Chapter 2). He used an idealised initial pressure field defined by a simple analytical
formula, with winds derived from it by means of the geostrophic relationship. He
then calculated the initial tendencies in the vicinity of England. On the basis of his
results, he drew conclusions about the limitations of geostrophic winds as initial
conditions. In this chapter we will re-examine his forecast using a global barotropic
model, and reconsider his conclusions.

Richardson's step-by-step description of his calculations in Chapter 2 of *WPNP* is
clear and explicit and serves as an excellent introduction to the process of numerical
weather prediction. Platzman (1967) wrote that he found the illustrative example
'one of the most interesting parts of the book'. In contrast to the straightforward style
of this chapter, much of the remainder of *WPNP* is heavy going, with the central
ideas often obscured by extraneous material. It appears likely that W. H. Dines
advised Richardson to add the introductory example, in the hope that it would make
the key ideas of numerical prediction more accessible to readers. Richardson ends
the chapter with an explicit acknowledgement: 'I am indebted to Mr W. H. Dines,
F.R.S., for having read and criticised the manuscript of this chapter.'

4.1 Richardson's model and data

The equations solved by Richardson were obtained by vertical integration of the
primitive equations linearised about an isothermal, motionless basic state. We recall

63

from (3.9) on page 49 that the vertical structure in this case is $Z = \exp(-z/\gamma H)$, so that vertical integration is equivalent to multiplication by the equivalent depth $h = \gamma H$. To facilitate comparison with Richardson's treatment, we let U, V, P denote the vertically integrated momenta and pressure. The equations (3.10)–(3.12) are formally unchanged by the integration:

$$\frac{\partial U}{\partial t} - fV + \frac{\partial P}{\partial x} = 0 \tag{4.1}$$

$$\frac{\partial V}{\partial t} + fU + \frac{\partial P}{\partial y} = 0 \tag{4.2}$$

$$\frac{\partial P}{\partial t} + gh\nabla \cdot \mathbf{V} = 0. \tag{4.3}$$

The dependent variables are the vertically integrated horizontal momentum $\mathbf{V} = (U, V) = (h\bar{\rho}u, h\bar{\rho}v)$ and the vertically scaled pressure perturbation $P = hp_S$ where p_S is the surface pressure. The independent variables are the time t, latitude ϕ and longitude λ (distances eastward and northward on the globe are given by $dx = a\cos\phi\,d\lambda$ and $dy = a\,d\phi$), a is the Earth's radius, g is the gravitational acceleration and $f = 2\Omega\sin\phi$ is the Coriolis parameter. The scale height $h = \gamma H = \gamma\Re T_0/g$ was set at 9.2 km, corresponding to a mean temperature $T_0 = 224$ K (Richardson denoted h by H' and based this value on observations made by his colleague, W. H. Dines). The reference density $\bar{\rho}$ may be determined by fixing the mean surface pressure; taking $p_0 = 1000$ hPa it is approximately 1.1 kg m^{-3}. Richardson used the C.G.S. system of units; the momentum values on page 8 of *WPNP* are approximately equivalent to the SI values of wind speed multiplied by 10^5; his pressure values are numerically equal to pressure in microbars or deci-Pascals.

The equations (4.1)–(4.3) are mathematically isomorphic to the linear shallow water equations that govern the small amplitude dynamics of an incompressible homogeneous shallow fluid layer on a sphere. In this case one interprets h as the mean depth and equates p_S to $\bar{\rho}\Phi'$ where $\Phi' = gz'$ is the geopotential perturbation of the free surface. The solutions of the Laplace tidal equations were discussed, using both analytical and numerical methods, in the previous chapter.

The initial pressure perturbation chosen by Richardson is depicted in Fig. 4.1. It is a simple zonal wavenumber one perturbation given by

$$p_S = 10^4 \sin^2\phi\cos\phi\sin\lambda \quad \text{(Pascals)} \tag{4.4}$$

which is symmetric about the equator with maxima of magnitude 38.5 hPa at 90°E and 55°N and S, and corresponding minima in the western hemisphere at the antipodes of the maxima. The initial winds were taken to be in geostrophic balance with the pressure field and the total column momenta are given (in kg m^{-1} s^{-1})

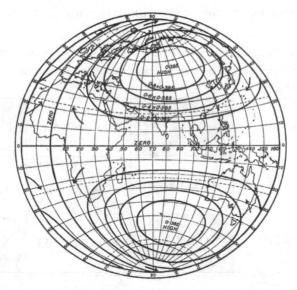

Figure 4.1 The initial pressure field chosen by Richardson for his barotropic forecast. (Reproduced from *WPNP*, p. 6)

by

$$U = -10^4 \frac{h}{2\Omega a}(2\cos^2\phi - \sin^2\phi)\sin\lambda, \quad V = +10^4 \frac{h}{2\Omega a}\sin\phi\cos\lambda. \quad (4.5)$$

The maximum velocity is about 20 m s^{-1} zonally along the equator. Figure 4.1 is an 'elevation' view of the eastern hemisphere on what appears to be a globular or equidistant projection. It is reproduced from page 6 of Richardson's book and was also featured on the cover of the Dover edition of *WPNP*.

4.2 The finite difference scheme

Richardson calculated the initial changes in pressure and wind (or momentum) using a finite difference method. He selected a time step of 2700 seconds or $\frac{3}{4}$ h, and used a spatial grid analogous to a chessboard, with pressure and momentum evaluated at the centres of the black and white squares respectively. The frontispiece of *WPNP*, reproduced in Fig. 1.6 on page 21, depicts such a grid, showing the chessboard pattern of alternating black and white squares. In this illustration, Richardson displaced the grid by two degrees west from that used in his introductory example, a subterfuge designed so that the grid points would coincide more closely with the actual distribution of observing stations in the vicinity of the United Kingdom.

A grid with the staggered discretisation of points proposed by Richardson is today called an E-grid, following the classification introduced by Arakawa (Arakawa and

Table 4.1 *Initial momentum values required to calculate the tendency at*
$\phi = 50.4°, \lambda = 0°.$

These values are extracted from the table on page 8 of *WPNP* and converted to SI units.

ϕ	$\cos\phi$	$\lambda_W = -\Delta\lambda$	$\lambda_0 = 0$	$\lambda_E = \Delta\lambda$
$\phi_N = 52.2°$	0.6129	P	$V_N = 78295.45$	P
$\phi_0 = 50.4°$	0.6374	$U_W = 1064.44$	P	$U_E = -1064.44$
$\phi_S = 48.6°$	0.6613	P	$V_S = 74327.53$	P

Lamb, 1977). Platzman suggested the name 'Richardson grid' and we will adopt that name. The distance between adjacent squares of like colour is 400 km in latitude, with 64 such squares around each parallel. This implies grid intervals of $2\Delta\lambda = 5.625°$, $2\Delta\phi = 3.6°$. The resolution at 50°N is about the same in both directions: the grid cells are approximately squares of side 200 km. A power of 2 was selected by Richardson for the number of longitudes, to permit repeated halvings of resolution towards the poles, where the meridians converge. A spherical grid system having almost homogeneous grid density over the globe was later proposed by Kurihara (1965), who made reference to *WPNP*, and the current European Centre for Medium-Range Weather Forecasts (ECMWF) model uses a reduced Gaussian grid, with a decreasing number of points around each parallel of latitude as the poles are approached.

The advantage of the chessboard pattern is that the time-rates are given at the points where the variables are initially tabulated. This is illustrated by consideration of the discretisation of the continuity equation.[1] Pressure is evaluated at the centres of the black squares and winds at the centres of the white. We write (4.3) in finite difference form as:

$$\frac{\partial P}{\partial t} + \frac{gh}{a\cos\phi}\left(\frac{\Delta U}{\Delta\lambda} + \frac{\Delta(V\cos\phi)}{\Delta\phi}\right) = 0. \tag{4.6}$$

Thus, to compute the tendency, we need the divergence on the black squares. For this, we require the zonal wind in the white squares to the east and west, and the meridional wind in the white squares to the north and south of each black square.

---

[1] The time discretisation will be discussed in §4.4 and in Chapter 5.

These are provided by the gridded values. A selection of values is given in Table 4.1. This is extracted from Richardson's Table on page 8 of *WPNP* (we convert his C.G.S. units to SI units by division by ten). The spatial grid increments are $\Delta\lambda = 0.0982$ radians and $\Delta\phi = 0.0628$ radians. There are two terms in the divergence:

$$\frac{\Delta U}{\Delta\lambda} = \frac{U_E - U_W}{\Delta\lambda} = -2.168 \times 10^4$$

$$\frac{\Delta(V\cos\phi)}{\Delta\phi} = \frac{V_N \cos\phi_N - V_S \cos\phi_S}{\Delta\phi} = -1.856 \times 10^4.$$

We note that they are both negative; there is no tendency for cancellation between them. The momentum divergence is given by

$$\nabla\cdot\mathbf{V} = \frac{1}{a\cos\phi_0}\left(\frac{\Delta U}{\Delta\lambda} + \frac{\Delta V\cos\phi}{\Delta\phi}\right) = -9.916 \times 10^{-3}$$

($a = 2 \times 10^7/\pi = 6.366 \times 10^6$ m) and the surface pressure tendency follows immediately:

$$\frac{\partial p_S}{\partial t} = -g\nabla\cdot\mathbf{V} \approx -9.79 \times (-9.916 \times 10^{-3}) = 9.708 \times 10^{-2}\,\mathrm{Pa\,s^{-1}}.$$

This corresponds to a change of pressure over the time step $\Delta t = \frac{3}{4}$ h of 2.621 hPa, in complete agreement with *WPNP*. This change may now be added to the initial pressure to obtain the pressure *at the same position* $\frac{3}{4}$ h later.

In a similar manner, considering the momentum equations (4.1) and (4.2), the grid structure ensures that the values required to calculate the Coriolis term and the pressure gradient term are provided at precisely those points where the momenta themselves are specified – the white cells – so the updating yields values at those points. Thus, the integration process can be continued without limit and is, in Richardson's words, 'lattice reproducing'. The chessboard pattern is depicted in Fig. 4.2 (from Platzman, 1967). The Richardson grid is composed of two C-grids, one comprising the quantities marked by asterisks, the other comprising the remaining quantities.

In replicating Richardson's results, there was some difficulty in obtaining complete agreement. The discrepancy was finally traced to Richardson's definition of the Earth's rotation rate. He specifies the value $2\Omega = 1.458423 \times 10^{-4}\,\mathrm{s^{-1}}$ on p. 13 of *WPNP*. This is correct. The *lazy man's* value $2\Omega_0 = 1.454441 \times 10^{-4}\,\mathrm{s^{-1}}$ is frequently used. It is defined using $\Omega_0 = 2\pi/(24 \times 60 \times 60)$, which implies a rotation in a solar day. In fact, the Earth makes one revolution in a sidereal day, giving Ω a slightly higher value. It was characteristic of Richardson to use the more accurate value.

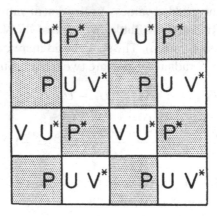

Figure 4.2 The Richardson grid. Pressure is evaluated at the centres of the black squares and winds at the centres of the white. The Richardson grid is composed of two C-grids, one comprising the quantities marked by asterisks, the other comprising the remaining quantities. (From Platzman (1967), © Amer. Met. Soc.)

4.3 Richardson's conclusions

The changes in pressure and momentum were calculated by Richardson for a selection of points near England and entered in the table on page 8 of *WPNP*. The focus here will be on the pressure tendency at the gridpoint 5600 km north of the equator and on the prime meridian (50.4°N, 0°E). The value obtained by Richardson was a change of 2.621 hPa in $\frac{3}{4}$ hour. This rise suggests a westward movement of the pressure pattern (similar rises were found at adjacent points). Such a movement is confirmed by use of the geostrophic wind relation in the pressure tendency equation (4.3). The momentum divergence corresponding to geostrophic flow is

$$\nabla \cdot \mathbf{V}_g = -\frac{\beta}{f} V_g = -\frac{\beta}{f^2} \frac{\partial P}{\partial x}. \tag{4.7}$$

This implies convergence for poleward flow and divergence for equatorward flow. Since the initial winds are geostropic, the initial tendency determined from (4.3) is

$$\frac{\partial p_S}{\partial t} = \frac{gh\beta}{f^2} \frac{\partial p_S}{\partial x} \tag{4.8}$$

(*WPNP*, Chapter 2#8). On the basis of his calculated tendency, Richardson comments that 'The deduction of this equation has been abundantly verified. It means that where pressure increases towards the east, there the pressure is rising if the wind be momentarily geostrophic' (*WPNP*, p. 9). Richardson argues that this deduction indicates that the geostrophic wind is inadequate for the computation of pressure changes.

The right-hand side of (4.8) is easily calculated for the chosen pressure perturbation (4.4). The parameter values are $a = 2 \times 10^7/\pi$ m, $g = 9.79\,\text{m s}^{-2}$, $h = 9.2\,\text{km}$, $2\Omega = 1.458 \times 10^{-4}\,\text{s}^{-1}$, in agreement with *WPNP*, p. 13. At the point in question the tendency calculated from (4.8) comes to $0.09713\,\text{Pa s}^{-1}$ or $2.623\,\text{hPa}$ in $\frac{3}{4}$ h. This analytical value is very close to the numerical value ($2.621\,\text{hPa}$) calculated by Richardson, indicating that errors due to spatial discretisation are small. The assumption of geostrophy implies zero initial tendencies for the velocity components; this follows immediately from (4.1) and (4.2). The calculated initial changes of momentum obtained by Richardson were indeed very small, confirming the accuracy of his numerical technique.

Richardson, pointing out that 'actual cyclones move eastward', described a result of observational analysis that showed a negative correlation between $\partial p_S/\partial t$ and $\partial p_S/\partial x$, in direct conflict with (4.8). In fact, the generally eastward movement of pressure disturbances in middle latitudes is linked to the prevailing strong westerly flow in the upper troposphere. Presumably, Richardson was unaware of the dominant influence of the jet stream, whose importance became evident only with the development of commercial aviation. However, he certainly was aware of the prevailing westerly flow in middle latitudes, and with its increase in intensity with height. Indeed, on page 7 of *WPNP*, he refers to Egnell's Law, an empirical rule stating that the horizontal momentum is substantially constant with height in the tropopause. This implies an increase in wind speed by a factor of roughly three between low levels and the tropopause.

Richardson was free to choose independent initial fields of pressure and wind, but 'it has been thought to be more interesting to sacrifice the arbitrariness in order to test our familiar idea, the geostrophic wind, by assuming it initially and watching the ensuing changes' (*WPNP*, p. 5). Recall that he worked out the introductory example in 1919, after his baroclinic forecast, the failure of which he ascribed to erroneous initial winds. He wanted to examine the efficacy of geostrophic winds and, moreover, Equation (4.8) which follows from the geostrophic assumption allowed him to compare the results of his numerical process with the analytical solution for the initial tendency. As has been seen above, the numerical errors were negligible. However, the implication of (4.8) for the westward movement of pressure disturbances led him to infer that the geostrophic wind is inadequate for the calculation of pressure changes. Towards the end of the chapter he concludes: 'It has been made abundantly clear that a geostrophic wind behaving in accordance with the linear equations ... [(4.1)–(4.3)] cannot serve as an illustration of a cyclone.' He then questions whether the inadequacy resides in the equations or in the initial geostrophic wind, or in both, and suggests that further analysis of weather maps is indicated.

Was Richardson's conclusion on the inappropriateness of geostrophic initial conditions justified? It seems from a historical perspective that the reason that his simple forecast lacked realism lies elsewhere. The zonal component of the geostrophic relationship is obtained by omitting the time derivative in (4.2). Taking the vertical derivative of this, after division by $h\bar{\rho}$, and using the hydrostatic equation and the logarithmic derivative of the equation of state one obtains the thermal wind equation:

$$\frac{\partial u_g}{\partial z} = -\left(\frac{g}{fT}\right)\frac{\partial T}{\partial y} + \left(\frac{u_g}{T}\right)\frac{\partial T}{\partial z}. \tag{4.9}$$

The second right-hand term may be neglected by comparison with the first. Taking a modest value of 30 K for the pole to equator surface temperature difference, and assuming that this is representative of the meridional temperature gradient in the troposphere, (4.9) implies a westerly shear of about $10\,\mathrm{m\,s^{-1}}$ between the surface and the tropopause. Thus, if the winds near the surface are light, there must be a marked westerly flow aloft. Although the importance of the jet stream was unknown in Richardson's day, the above line of reasoning was obviously available to him. Yet, he focused on the geostrophic wind as the cause of lack of agreement between his forecast and the observed behaviour of cyclones in middle latitudes. The real reason for this was his assumption of a barotropic structure for his disturbance, in contrast to the profoundly baroclinic nature of extratropical depressions.

Richardson might have incorporated the bulk effect of the zonal flow on his perturbation by linearising the equations about a westerly wind, $\bar{u}$. This would result in a modification of (4.8) to

$$\frac{\partial p_S}{\partial t} = \left(\frac{gh\beta}{f^2} - \bar{u}\right)\frac{\partial p_S}{\partial x}.$$

It would appear that, for sufficiently strong mean zonal flow, the sign of the pressure tendency would be reversed. However, things are not quite so simple. The effect of a background zonal flow is not a simple Doppler shift. The meridional temperature gradient that appears in the lower boundary condition tends to counteract the Doppler shift of the zonal flow. This 'non-Doppler effect' was discussed by Lindzen, 1968 (see also Lynch, 1979).

4.4 The global numerical model

Let us now consider the results of repeating and extending Richardson's forecast using a global barotropic numerical model based on the linear equations used by him, the Laplace tidal equations. Precisely the same initial data were used, and the same time step and spatial grid interval were chosen. There were only two substantive differences from Richardson's method. First, an implicit time-stepping scheme was used to insure numerical stability, in contrast to the explicit scheme of *WPNP* which is restricted by the Courant–Friedrichs–Lewy criterion (see §5.3 below). Second, a C-grid was used for the horizontal discretisation. The C-grid has zonal wind represented half a grid-step east, and meridional wind half a grid-step north of the pressure points. This grid was chosen when the model was first designed, for reasons that are irrelevant in the current context. Richardson's grid comprises two C-grids superimposed, and coupled only through the Coriolis terms (Fig. 4.2). The initial pressure tendency calculated on one C-grid is independent of the other one.

The numerical scheme used to solve the shallow water equations is a linear version of the two-time-level scheme devised by McDonald (1986). The momentum equations (4.1) and (4.2) are integrated in two half-steps $\frac{1}{2}\Delta t$. In the first half-step, the Coriolis terms are implicit and the pressure terms explicit:

$$\left(U^{n+\frac{1}{2}} - U^n\right) = \tfrac{1}{2}\Delta t\left(f V^{n+\frac{1}{2}} - P_x^n\right),$$
$$\left(V^{n+\frac{1}{2}} - V^n\right) = \tfrac{1}{2}\Delta t\left(- f U^{n+\frac{1}{2}} - P_y^n\right).$$

In the second half-step, the Coriolis terms are explicit and the pressure terms implicit:

$$\left(U^{n+1} - U^{n+\frac{1}{2}}\right) = \tfrac{1}{2}\Delta t\left(f V^{n+\frac{1}{2}} - P_x^{n+1}\right),$$
$$\left(V^{n+1} - V^{n+\frac{1}{2}}\right) = \tfrac{1}{2}\Delta t\left(- f U^{n+\frac{1}{2}} - P_y^{n+1}\right).$$

The Coriolis terms are said to be pseudo-implicit, because the intermediate values $U^{n+\frac{1}{2}}$ and $V^{n+\frac{1}{2}}$ can be eliminated immediately, yielding

$$U^{n+1} + \tfrac{1}{2}\Delta t\, P_x^{n+1} = R_U, \tag{4.10}$$
$$V^{n+1} + \tfrac{1}{2}\Delta t\, P_y^{n+1} = R_V, \tag{4.11}$$

where the right-hand terms $\mathbf{R} = (R_U, R_V)$ depend only on the variables at time $n\Delta t$. An expression for the divergence follows immediately:

$$\nabla \cdot \mathbf{V}^{n+1} + \tfrac{1}{2}\Delta t\, \nabla^2 P^{n+1} = \nabla \cdot \mathbf{R}. \tag{4.12}$$

The continuity equation (4.3) is integrated in a single centred implicit step Δt:

$$(P^{n+1} - P^n) = \tfrac{1}{2}\Delta t\, gh\left(\nabla \cdot \mathbf{V}^{n+1} + \nabla \cdot \mathbf{V}^n\right). \tag{4.13}$$

Elimination of the divergence between (4.12) and (4.13) yields an equation for the pressure:

$$[\nabla^2 - K^2]P^{n+1} = R_P, \tag{4.14}$$

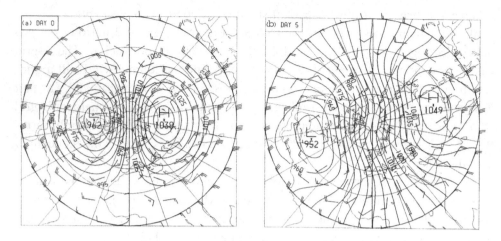

Figure 4.3 (a) The initial pressure and wind fields in the Northern Hemisphere as specified by Richardson. (b) Forecast pressure and wind fields valid five days later. (From Lynch (1992), © Amer. Met. Soc.)

where $K^2 = 4/(\Delta t^2 gh)$ and the right-hand forcing R_P is known from variables at time $n\Delta t$. This Helmholtz equation is solved by the method of Sweet (1977). Once P^{n+1} is known, U^{n+1} and V^{n+1} follow from (4.10) and (4.11). The simple form of (4.14) is a consequence of splitting the integration of the momentum equations into two half-steps. However, this splitting is formal: no $(n + \frac{1}{2})\Delta t$ values are actually computed. An analysis by McDonald (1986) showed the scheme to be unconditionally stable and to have time truncation $O(\Delta t^2)$.

4.5 Extending the forecast

The pressure and wind selected by Richardson and specified by (4.4) and (4.5) are shown in Fig. 4.3(a). The forecast fields of pressure and wind valid five days later may be seen in Fig. 4.3(b). Only the Northern Hemisphere is shown, as the fields south of the equator are a mirror image of those shown. The initial change in pressure at the location 50.4°N, 0°E was 2.601 hPa in $\frac{3}{4}$ hour. This is close to Richardson's value (2.621 hPa), the slight difference being due to the implicit time-stepping scheme. The wavenumber one perturbation rotates westward almost circumnavigating the globe in five days. The pattern is substantially unchanged in extratropical regions. In the tropics a zonal pressure gradient, absent in the initial field, develops and is maintained thereafter.

One may ask why the chosen perturbation is so robust, maintaining a coherent structure for such a long time and undergoing little change save for the retrogressive or westward rotation. This behaviour is attributable to the close resemblance of

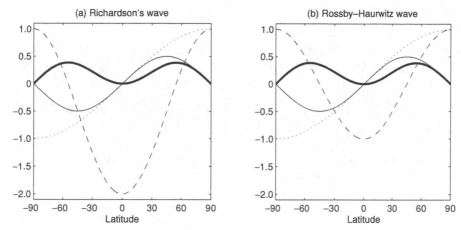

Figure 4.4 (a) Richardson's initial conditions. (b) First symmetric Rossby–Haurwitz wave RH(1,2). Dashed lines u, dotted lines v, solid lines P. The pressures and meridional winds are identical in (a) and (b); only the zonal winds differ.

Richardson's chosen data to an eigensolution of the linear equations. The normal-mode solutions of the Laplace tidal equations were studied in the previous chapter and it was found that the solutions fall into two categories, the gravity-inertia waves and the rotational modes. Haurwitz (1940) found that, if divergence is neglected, the frequencies of the rotational modes are given by the dispersion relation

$$\nu = -\frac{2\Omega m}{n(n+1)} \tag{4.15}$$

where m and n are the zonal and total wavenumbers. For $m = 1$ and $n = 2$ this gives a period of three days. The stream function of the Rossby–Haurwitz (RH)(1,2) mode is

$$\psi = \psi_0 \Im\{Y_2^1(\lambda, \phi)\} = \psi_0 \sin\phi\cos\phi\sin\lambda \tag{4.16}$$

and we notice that this is related to Richardson's choice of pressure through the simplified linear balance relationship $p_S = f\bar{\rho}\psi$. The wind $\mathbf{v}_\psi = \mathbf{k} \times \nabla\psi$ is, of course, nondivergent. But the geostrophic wind $\mathbf{v}_g$ derived from p_S is not. The two are related by

$$\mathbf{v}_g = \mathbf{v}_\psi - (\beta\psi/f)\mathbf{i}$$

where $\mathbf{i}$ is an eastward-pointing unit vector. Thus, the zonal wind chosen by Richardson differs from that of RH(1,2) and gives rise to non-vanishing divergence. The pressure and meridional wind components are identical: the structures of the two waves are shown in Fig. 4.4. The period of RH(1,2) is three days but, as we saw in the previous chapter, the period of the corresponding Hough mode is approximately five days.

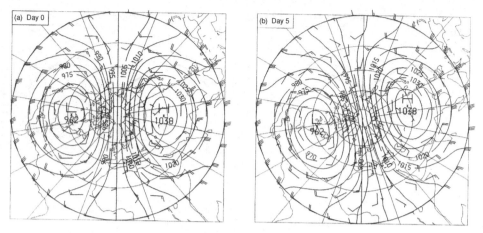

Figure 4.5 The five-day wave, the gravest symmetric rotational eigenfunction. (a) The initial pressure and wind fields in the Northern Hemisphere. (b) Forecast pressure and wind fields valid five days later. The wave has almost completed a circuit of the globe. (From Lynch (1992), © Amer. Met. Soc.)

The horizontal structure of the five-day wave is shown in Fig. 4.5(a) (with maximum amplitude of 38.5 hPa as for Richardson's field). The global model was integrated for five days using this wave as initial data and the result is shown in Fig. 4.5(b). It is practically indistinguishable from the initial field except for the angular displacement (it has almost completed a full circuit of the globe), consistent with the expected behaviour of an eigenfunction. The initial pressure tendency calculated for the five-day wave at the gridpoint at 50.4°N, 0°E was 1.39 hPa in $\frac{3}{4}$ h. Recall that Richardson's data produced an initial change of 2.62 hPa. Why the difference? Suppose that the pressure pattern (4.4) defined by Richardson were to rotate westward without change of form, making one circuit of the Earth in a period $\tau = 5$ days. It could then be described by the equation

$$p_{\mathrm{S}} = 10^4 \sin^2 \phi \cos \phi \sin(\lambda + 2\pi t/\tau) \qquad (4.17)$$

and the rate of change is obviously

$$\frac{\partial p_{\mathrm{S}}}{\partial t} = \left(\frac{2\pi}{\tau}\right) 10^4 \sin^2 \phi \cos \phi \cos(\lambda + 2\pi t/\tau) \qquad (4.18)$$

which, at the point 50.4°N, 0°E at time $t = 0$, gives the value 1.44 hPa in $\frac{3}{4}$ h, about half the value obtained by Richardson. One must conclude that the evolution of his pressure field involves more than a simple westward rotation with period τ.

An analysis of Richardson's initial data into Hough mode components was carried out by Lynch (1992). It was shown that almost 85 per cent of the energy projects onto the five-day wave and about 11 per cent onto the Kelvin wave. Over 99 per cent of the energy is attributable to just three Hough mode components. Although relatively little energy resides in the gravity wave components, they contribute to the tendency due to their much higher frequencies. The temporal variations can be accounted for by the primary component in each of the three categories: rotational, eastward gravity and westward gravity waves. The normal mode analysis of the fields used for the barotropic forecast indicated the presence of a westward travelling gravity wave with a period of about 13.5 hours. This component could be clearly seen in the time-traces of the forecast fields (Lynch, 1992, Fig. 8). From a synoptic viewpoint, the short-period variations due to the gravity wave components may be considered as undesirable noise. Such noise may be removed by modification of the initial data, a process known as initialisation.

A simple initialisation technique, using a digital filter, was applied to the initial fields, and was successful in annihilating the high frequency oscillations without significantly affecting the evolution of the longer period components (Lynch, 1992). Although the barotropic model is linear, this initialisation technique may also be used with a fully nonlinear model. The baroclinic forecast which forms the centrepiece of Richardson's book was spoiled by high frequency gravity wave components in the initial data. These gave rise to spuriously large values of the pressure tendency. Since the design of the low-pass filter and its application to the initialisation of Richardson's multi-level forecast will be described in a later chapter, we will not discuss it further here.

4.6 Non-divergent and balanced initial conditions

Richardson recognised that the divergence field calculated from his data was unrealistic, and he blamed this for the failure of the baroclinic forecast. However, even if the wind fields were modified to remove divergence completely, high-frequency gravity waves would still be present, and cause spurious oscillations. We will consider the baroclinic forecast in depth later. For the present, we investigate the effectiveness of using nondivergent initial winds for the single-level forecast. Two 24-hour forecasts were carried out: one started from Richardson's data as specified in (4.4) and (4.5); the other started from the same pressure field (4.4) but with wind fields derived from the stream function (4.16). The pressure is related to ψ by the linear balance relationship $p_S = f \bar{\rho} \psi$. The latter conditions correspond to the Rossby–Haurwitz wave RH(1,2). The latitudinal structure of the two sets of initial conditions was shown in Fig. 4.4.

The pressure and meridional wind component are identical for the two initial states; only the zonal winds differ. The level of high-frequency noise can be estimated by calculating the mean absolute pressure tendency. This is equivalent, through the continuity equation, to the mean absolute divergence, which we denote

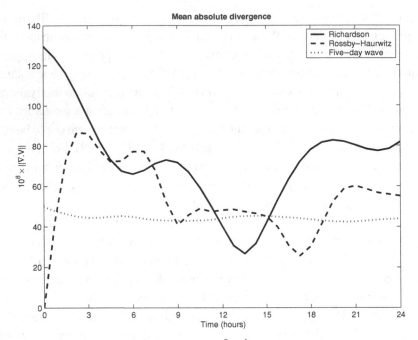

Figure 4.6 Mean absolute divergence ($\times 10^8$ s^{-1}) for three 24-hour forecasts. Solid line: Richardson's initial conditions. Dashed line: non-divergent initial conditions (Rossby–Haurwitz mode, $m = 1$, $n = 2$). Dotted line: pure Hough mode initial data (five-day wave).

by N_1. The evolution of N_1 for the two forecasts is shown in Fig. 4.6. The solid line is for the forecast from Richardson's initial fields. The dashed line is for a forecast starting with the Rossby–Haurwitz structure. Obviously, the divergence for the RH(1,2) conditions vanishes at the initial time. However, it does not remain zero: the RH wave is a solution of the non-divergent barotropic vorticity equation, but it is not an eigensolution of the Laplace tidal equations. We see that after about three hours the divergence has grown from zero to a value comparable to that for Richardson's data. Thus, setting the divergence to zero at the initial time is not an effective means of ensuring that it remains small.

To demonstrate the beneficial effect of using properly balanced initial conditions, a third forecast was done, with the five-day wave specified as initial data. The amplitude was scaled so that the maximum pressure was the same for all three forecasts. The horizontal structure of these initial conditions was shown in Fig. 4.5. The five-day wave is a pure rotational mode, so that there is no projection onto the high-frequency gravity waves. The evolution of the noise parameter N_1 for the third forecast is shown by the dotted line in Fig. 4.6. We see that it remains more-or-less constant throughout the 24-hour forecast, and at a level somewhat less than that of the other two forecasts (if numerical errors were completely absent, it would remain perfectly constant). This confirms that it is possible to choose initial data for which the ensuing forecast is free from spurious noise, but that *a non-vanishing divergence field is required for balance*. This issue will be examined in greater detail in Chapter 8, when normal mode initialisation will be considered.

4.7 Reflections on the single layer model

With hindsight, we can understand Richardson's introductory example as a remarkable simulation of the 'five-day wave'. This wave, the gravest symmetric rotational normal mode of the atmosphere, has been unequivocally detected in atmospheric data (Madden and Julian, 1972; Madden, 1979). It is almost always present, is global in extent with maximum amplitude in middle latitudes, extends at least to the upper stratosphere and has no phase-shift with either latitude or height. Geisler and Dickinson (1976) modelled the five-day wave in the presence of realistic zonal winds and found that its period was remarkably robust, being little affected by the background state because the zonal flow and temperature gradient at the Earth's surface produce almost equal but opposite changes in period. Rodgers (1976) found clear evidence for the five-day wave in the upper stratosphere.

Richardson regarded the westward movement of his pressure pattern to be in conflict with the generally observed eastward progression of synoptic systems in middle latitudes, and concluded that '... use of the geostrophic hypothesis was found to lead to pressure-changes having an unnatural sign' (*WPNP*, p. 146). He also neglected the dominant influence of the westerlies in the upper troposphere. Although the jet stream was yet to be discovered, the existence of westerly shear follows from simple balance considerations. Ironically, had Richardson included a realistic zonal flow component in his introductory example, he would still have found that the pressure disturbance moved westward, due to the robustness of its period (Lynch, 1979).

Richardson was unaware of the importance of the five-day wave and other such free waves in the atmosphere. Although these solutions had been studied much earlier by Margules and Hough, their geophysical significance became apparent only with the work of Rossby (1939). Richardson made reference to Lamb's *Hydrodynamics*, where tidal theory was reviewed, but he considered (*WPNP*, p. 5) that 'its interest has centred mainly in forced and free oscillations, whereas now we are concerned with unsteady circulations'. He was convinced that the single layer barotropic model was incapable of providing a useful representation of atmospheric dynamics.

Richardson and Munday (1926) carried out a thorough and meticulous investigation of data from sounding balloons to examine the 'single layer' question. They showed, in particular, that if the pressure were everywhere a function of density only, then the atmosphere would behave as a single layer. The term *barotropic* had been introduced by Vilhelm Bjerknes (1921) to describe such a hypothetical state. Richardson found nothing in the observations to justify the single layer assumption and concluded that Laplace's tidal equations were 'a very bad description of ordinary disturbances of the European atmosphere'. The values of the dynamical height h estimated from the data were extremely variable and an assumption of constant h could not be justified. In the Revision File, inserted before page 23, there is a copy of a letter from Richardson to Harold Jeffreys, dated 20 April 1933, where he writes 'The concept of a dynamical height of the atmosphere is a very great idealisation. . . . Any argument based on the fixity of h is rotten.'

It is unfortunate that Richardson argued so pursuasively against the utility of the single-layer model. As we will see, a simple barotropic model proved to be of great practical value during the early years of operational numerical weather prediction. Richardson's conclusions pertained only to the linear tidal equations. The advective process, embodied in the nonlinear terms, was not considered. Moreover, the process of group velocity, whereby the energy of a disturbance can propagate eastward while individual wave components proceed westward, had yet to be elucidated in the context of atmospheric dynamics (Rossby, 1945). Richardson's lack of confidence in the barotropic representation of the atmosphere was shared by many of his colleagues, with the result that operational numerical weather prediction in Britain was delayed until baroclinic effects could be successfully modelled. We will discuss this in more detail in Chapter 10.

5

The solution algorithm

The procedure to be described is certainly very complicated, but so are the atmospheric changes.   (*WPNP*, p. 156)

Richardson's numerical processes, which we began to study in the context of the barotropic model, will be considered in greater detail now. Richardson realised that analytical solutions of the nonlinear equations could not be obtained. He therefore turned to numerical methods and developed approximate solution procedures, some of which are still in common use. For example, his *deferred approach to the limit* (Richardson, 1927) is at the basis of the Bulirsch–Stoer method used in a currently popular software package (Press *et al.*, 1992). Marchuk (1982) devoted an entire chapter of his book to this method. A brief review of Richardson's work in numerical analysis was included in the *Collected Papers* (Fox, 1993).

In this chapter we review the numerical analysis required for an understanding of Richardson's procedure, paying special attention to the leapfrog method that he employed. We investigate the accuracy of finite difference approximations and consider the important question of computational stability. (Much greater detail on numerical methods for atmospheric models is available in texts such as Durran (1999), Haltiner and Williams (1980), Mesinger and Arakawa (1976) and Kalnay (2003)). A detailed step-by-step description of Richardson's solution procedure will be presented. His systematic algorithm is, in many ways, similar in structure to numerical weather prediction models in use today.

5.1 The finite difference method

The numerical solution of the differential equations that govern the evolution of the atmosphere requires that they be transformed into algebraic form. This is done by the now-familiar method of finite differences: the continuous variables are represented by their values at a finite set of points, and derivatives are approximated by differences between values at adjacent points. The atmosphere is partitioned into several discrete horizontal layers, each layer is divided up into a number of grid cells, and the variables are evaluated at the centre of each cell.[1] Similarly, the time interval under consideration is sliced into a finite number of discrete time steps, and the continuous evolution of the variables is approximated by the change from step to step.

The finite difference method was developed by Richardson in his study of stresses in masonry dams (Richardson, 1910). In a sense, the method corresponds to a reversal of history. The derivative df/dx of a function $f(x)$ exists if the ratio

$$\frac{\Delta f}{\Delta x} = \frac{f(x + \Delta x) - f(x)}{\Delta x} \tag{5.1}$$

tends to a definite limit as the increment Δx tends to zero. This ratio is uncentred, or unsymmetric about the point x. We may also consider the centred difference ratio

$$\frac{\Delta f}{2\Delta x} = \frac{f(x + \Delta x) - f(x - \Delta x)}{2\Delta x}. \tag{5.2}$$

Differential calculus depends upon justifying the limiting process $\Delta x \to 0$. In approximating a differential equation by finite differences, we reverse the procedure, and replace derivatives by corresponding ratios of increments of the dependent and independent variables. Richardson described the procedure thus:

Although the infinitesimal calculus has been a splendid success, yet there remain problems in which it is cumbrous or unworkable. When such difficulties are encountered it may be well to return to the manner in which they did things before the calculus was invented, postponing the passage to the limit until after the problem has been solved for a moderate number of moderately small differences.   *(Richardson, 1927)*

The accuracy of the finite difference approximation is measured by the size of the residual

$$\varepsilon(x, \Delta x) = \left| \frac{d f}{dx} - \frac{\Delta f}{\Delta x} \right|,$$

which depends on Δx. Richardson was one of the first to analyse the behaviour of the errors. If the finite difference is centred about the point where the approximation is required, as in (5.2), the accuracy is typically of the second order; that is, the error diminishes like Δx^2 as $\Delta x \to 0$. Uncentred differences like (5.1) usually lead to difference schemes of only first-order accuracy.

Let us assume that the domain of x is represented by a sequence of discrete points $x_m = m\Delta x$ and the values of the function at these points are denoted by $f_m =$

---

[1] Richardson used the term 'lattice', borrowed from crystallography, for the discrete spatial grid.

$f(x_m)$. Let us substitute the exact function $f(x)$ into the one-sided or uncentred approximation (5.1) to the derivative, and expand in a Taylor series about x_m:

$$\frac{f_{m+1} - f_m}{\Delta x} = \left(\frac{d f}{dx}\right)_m + \frac{1}{2}\left(\frac{d^2 f}{dx^2}\right)_m \Delta x + \frac{1}{6}\left(\frac{d^3 f}{dx^3}\right)_m \Delta x^2 + \ldots .$$

The first term on the right is the exact derivative. The remaining terms comprise the error. We see that it varies in proportion to the grid-size: it is first-order, or $O(\Delta x)$. Now repeating the process using the centred approximation (5.2) for the derivative, we obtain

$$\frac{f_{m+1} - f_{m-1}}{2\Delta x} = \left(\frac{d f}{dx}\right)_m + \frac{1}{6}\left(\frac{d^3 f}{dx^3}\right)_m \Delta x^2 + O(\Delta x^4).$$

The truncation error is now $O(\Delta x^2)$, which is significantly more accurate. Richardson was careful to use centred differences wherever possible. In the opening chapter of *WPNP* he mentions two important properties of finite differences 'brought to notice by Mr W. F. Sheppard' (Sheppard, 1899):

(a) The great gain in accuracy, in the representation of a differential coefficient, when the differences are centred instead of progressive; a gain secured by a slight increase of work.
(b) The errors due to centred differences, which are proportional to the square of the co-ordinate difference. This fact provides a universal means of checking and correcting the errors.

5.2 Integration in time

Two distinct, special types of problem arise involving the solution of partial differential equations, i.e. boundary value problems and initial value problems. The solution of an elliptic equation on a limited spatial domain is a boundary value problem: the solution may be determined if the values (or gradients) of the unknown function are given at the boundary. Richardson called such problems 'jury problems'. The solution has to satisfy all the boundary conditions, 'just as the verdict has to satisfy all the jurymen sitting round the table' (Richardson, 1925). He noted that the solution of these problems could involve 'troublesome successive approximations' (*WPNP*, p. 3). The method of successive approximations that he had developed (Richardson, 1910) was designed to solve just such problems. By contrast, initial value problems are those for which the solution is determined once its value at a particular time is given. Richardson coined the term 'marching problems', as the solution is calculated step by step so that our knowledge of it marches forward one step at a time. We will see that, for the finite-difference solution of such problems to yield a good approximation to the solution of the continuous equations, there is a limitation, which was unknown to Richardson, on the maximum size of the time step. The equations determining atmospheric flow constitute a mixed initial-boundary problem but, before considering the complete system, it is useful to study the solution of pure initial value problems.

5.2.1 The leapfrog time scheme

We consider first the method of advancing the solution in time. It is performed by means of the now-popular leapfrog scheme, called by Richardson the *step-over method* (WPNP, p. 150). Let Q denote a typical dependent variable, governed by an equation of the form

$$\frac{dQ}{dt} = F(Q). \tag{5.3}$$

The continuous time domain t is replaced by $\{0, \Delta t, 2\Delta t, \ldots, n\Delta t, \ldots\}$, a sequence of discrete moments, and the solution at these moments is denoted by $Q^n = Q(n\Delta t)$. If this solution is known up to time $t = n\Delta t$, the right-hand term $F^n = F(Q^n)$ can be computed. The time derivative in (5.3) is now approximated by a centred difference

$$\frac{Q^{n+1} - Q^{n-1}}{2\Delta t} = F^n,$$

so the 'forecast' value Q^{n+1} may be computed from the old value Q^{n-1} and the tendency F^n:

$$Q^{n+1} = Q^{n-1} + 2\Delta t \, F^n. \tag{5.4}$$

This process of stepping forward from moment to moment is repeated a large number of times, until the desired forecast range is reached.

For the physical equation (5.3), a single initial condition Q^0 is sufficient to determine the solution. One problem with the leapfrog scheme (and other three-time-level schemes) is that two values of Q are required to start the computation. In addition to the *physical* initial condition Q^0, a *computational* initial condition Q^1 is required. This cannot be obtained using the leapfrog scheme, so a simple non-centred step

$$Q^1 = Q^0 + \Delta t \, F^0$$

is used to provide the value at $t = \Delta t$; from then on, the leapfrog scheme can be used. However, as noted by Richardson (WPNP, p. 151), the errors of the first step will persist.

5.2.2 Numerical stability of time schemes

The friction equation

In a section entitled *The Arrangement of Instants* (WPNP, §7/2), Richardson set down several methods of time-integration, and applied them to the simple *friction equation*:

$$\frac{dQ}{dt} = -\kappa Q, \quad \text{with} \quad Q = Q^0 \quad \text{at} \quad t = 0 \tag{5.5}$$

(he set $\kappa = 1$ and $Q^0 = 1$; we assume only $\kappa > 0$). Of course, the analytical solution is $Q(t) = Q^0 \exp(-\kappa t)$, which decays monotonically with time. The simplest numerical scheme is the Euler forward method, which Richardson called the method of advancing time-steps. The time derivative in the differential equation (5.5) is approximated by a forward difference:

$$\frac{Q^{n+1} - Q^n}{\Delta t} = -\kappa Q^n.$$

It is easy to find the solution of this difference equation: $Q^n = Q^0(1 - \kappa \Delta t)^n$. This solution decays monotonically in time provided $\kappa \Delta t < 1$. The Euler scheme is *stable* for this friction equation. However, the scheme is only first-order accurate: for a fixed time $\tau = n \Delta t$, the error is

$$\varepsilon(\tau) = |Q^0(1 - \kappa \Delta t)^n - Q^0 \exp(-\kappa n \Delta t)| = \tfrac{1}{2} Q^0 \tau \kappa^2 \Delta t + O(\Delta t^2).$$

We might attempt to obtain a more accurate solution by using a centred difference for the time derivative, as in the leapfrog scheme:

$$\frac{Q^{n+1} - Q^{n-1}}{2\Delta t} = -\kappa Q^n. \tag{5.6}$$

A solution in the form of a geometric progression, $Q^n = Q^0 \alpha^n$ for some constant α, will decrease monotonically provided $0 < \alpha < 1$. This is the condition for the character of the finite difference solution to resemble that of the continuous equation. Substituting this solution into (5.6), it follows that there are two possibilities:

$$\alpha_+ = -\kappa \Delta t + \sqrt{1 + \kappa^2 \Delta t^2} \quad \text{and} \quad \alpha_- = -\kappa \Delta t - \sqrt{1 + \kappa^2 \Delta t^2}.$$

It is easy to see that $|\alpha_+| < 1$ for all Δt so that a decaying solution is obtained. However, $|\alpha_-| > 1$ for all Δt so that this solution grows without limit with n. The first solution gives the *physical mode*, which is similar in character to the solution of the differential equation. The second solution is called the *computational mode*. It appears as a result of replacing a first-order derivative by a second-order difference. It has no counterpart in the physical equation but is a completely spurious numerical artifact. Since $\alpha_- < 0$, its sign alternates each time-step and, since it grows with time, it leads to numerical instability. Thus, despite its second-order accuracy, the leapfrog scheme produces unacceptable errors for the friction equation.

Richardson calculated the solution of the friction equation (5.5) using the leapfrog scheme and noted 'a curious oscillation in time ... the amplitude increasing as it goes'. As Platzman (1967, p. 527) has observed, Richardson was here on the verge of detecting the phenomenon of computational instability. The amplitude of the computational mode is determined by the initial conditions. Use of a smooth start can minimise its impact. Richardson considered several methods of starting the integration. The most practical is to use a very small uncentred step and double the length of the time-step repeatedly until the desired size of Δt is reached. He found that a careful start-up procedure greatly reduced the errors of the solution. However, the existence of the unstable numerical mode and the inevitable presence of numerical noise guarantee that the solution of the friction equation will ultimately become meaningless unless some additional controlling mechanism is introduced.

The oscillation equation

The instability of the leapfrog scheme would appear to make it unsuitable for use. However, if we apply it to the simple *oscillation equation*

$$\frac{dQ}{dt} = i\omega Q, \qquad \text{with} \quad Q = Q^0 \quad \text{at} \quad t = 0, \tag{5.7}$$

a markedly different result is obtained. The analytical solution is $Q(t) = Q^0 \exp(i\omega t)$. Using the leapfrog scheme and again seeking a solution $Q^n = Q^0 \alpha^n$ for constant α, there are again two possibilities:

$$\alpha_\pm = i\omega\Delta t \pm \sqrt{1 - \omega^2 \Delta t^2}.$$

For $|\omega\Delta t| \leq 1$, it is clear that $|\alpha_\pm| = 1$ so that two oscillating solutions are obtained. For small $\omega\Delta t$ we have $\alpha_+ \approx +1$ and $\alpha_- \approx -1$. The former approximates the analytical solution. The latter is the computational mode, which alternates in sign from step to step. While it can be troublesome, its amplitude will remain small if it is initially small; the leapfrog scheme is stable for the oscillation equation, provided $|\omega\Delta t| < 1$.

Mesinger and Arakawa (1976) used a simple example to illustrate the behaviour of the computational mode for the leapfrog scheme. They considered (5.7) with $\omega = 0$. The true solution is $Q = Q^0$, constant. The leapfrog scheme gives

$$Q^{n+1} = Q^{n-1}. \tag{5.8}$$

They consider two particular choices of Q^1. First, suppose the exact value $Q^1 = Q^0$ is chosen. Then the numerical solution of (5.8) is $Q^n = Q^0$ for all n, which is exact. Second, suppose $Q^1 = -Q^0$. The solution of (5.8) is $Q^n = (-1)^n Q^0$, which is comprised entirely of the computational mode. This illustrates the importance of a careful choice of the computational initial condition.

Mixed and implicit schemes

The leapfrog scheme is stable for the oscillation equation and unstable for the friction equation. The Euler forward scheme is stable for the friction equation but unstable for the oscillation equation. If we require an approximation to an equation such as

$$\frac{dQ}{dt} = i\omega Q - \kappa Q,$$

containing terms of both types, a reasonable approach might be to use the leapfrog scheme for the oscillation term and the forward scheme for the friction term:

$$Q^{n+1} = Q^{n-1} + 2\Delta t(i\omega Q^n - \kappa Q^{n-1}).$$

We may show that this is stable provided $(2\kappa \Delta t + \omega^2 \Delta t^2) \leq 1$. Modern numerical models of the atmosphere typically combine several distinct schemes in this way. They also use various filtering processes to limit spatial and temporal noise.

Richardson also considered the use of an implicit numerical method. For the simple oscillation equation (5.7), this amounts to the approximation

$$\frac{Q^{n+1} - Q^n}{\Delta t} = i\omega \left(\frac{Q^{n+1} + Q^n}{2} \right). \tag{5.9}$$

A straightforward analysis shows that this scheme is second-order accurate and unconditionally stable. In this simple case, we may solve immediately for Q^{n+1}. In general, this requires solution of a complicated nonlinear system. For that reason, it is common practice today to treat selected linear terms implicitly and the remaining terms explicitly. This 'semi-implicit' method was pioneered by André Robert (for a review, see Robert, 1979). The terms that give rise to high frequency gravity waves are integrated implicitly, enabling the use of a long time step. Richardson wrote that the implicit scheme was 'probably unworkable ... [for the meteorological equations], on account of the complexity of the system of equations'. Indeed, an implicit numerical scheme would have been quite unmanageable for Richardson's manual calculation, but schemes of this sort are pivotal in modern weather prediction models, due to their desirable stability properties.

5.3 The Courant–Friedrichs–Lewy stability criterion

So far, we have considered finite difference approximations to ordinary differential equations. For the partial differential equations that govern atmospheric dynamics, more subtle stability criteria apply. The accuracy depends on the sizes of both the space and time steps, Δx and Δt. We might reasonably expect that the errors should become smaller as the spatial and temporal grids are refined. However, a surprising new factor enters: the stability depends on the *relative* sizes of the space and time steps. A realistic solution is not guaranteed by reducing the sizes of Δx and Δt independently.

We consider the simple wave equation

$$\frac{\partial Q}{\partial t} + c\frac{\partial Q}{\partial x} = 0, \tag{5.10}$$

where $Q(x, t)$ depends on both x and t. The advection speed c is constant and, without loss of generality, we assume $c > 0$. The wave equation has a general solution of the form $Q = f(x - ct)$, where f is an arbitrary function. In particular, we may consider the sinusoidal solution $Q = Q^0 \exp[ik(x - ct)]$ of wavelength $L = 2\pi/k$. We use centred difference approximations in both space and time:

$$\left(\frac{Q_m^{n+1} - Q_m^{n-1}}{2\Delta t}\right) + c\left(\frac{Q_{m+1}^n - Q_{m-1}^n}{2\Delta x}\right) = 0, \tag{5.11}$$

where $Q_m^n = Q(m\Delta x, n\Delta t)$. We seek a solution of the form $Q_m^n = Q^0 \exp[ik(m\Delta x - Cn\Delta t)]$. For real C, this is a wave-like solution. However, if C is complex, this solution will behave exponentially, quite unlike the solution of the continuous equation. Substituting Q_m^n into the finite difference equation, we find that

$$C = \frac{1}{k\Delta x}\sin^{-1}\left[\left(\frac{c\Delta t}{\Delta x}\right)\sin k\Delta x\right]. \tag{5.12}$$

If the argument of the inverse sine is less than unity, C is real. Otherwise, C is complex, and the solution will grow with time. Clearly, $c\Delta t/\Delta x \le 1$ is a sufficient condition for real C. It is also necessary: for a wave of four gridlengths, $k\Delta x = \pi/2$, so $\sin k\Delta x = 1$. Thus, the condition for stability of the solution is

$$Le \equiv \frac{c\Delta t}{\Delta x} \le 1. \tag{5.13}$$

This non-dimensional parameter is often called the Courant number, but is denoted here as Le for Lewy, who first discovered this stability criterion (Reid, 1976, p. 116). It was published by Courant, Friedrichs and Lewy (Courant *et al.*, 1928) and is called the Courant–Friedrichs–Lewy (CFL) stability criterion. It imposes a strong constraint on the relative sizes of the space and time grids. It was of course unknown to Richardson, but John von Neumann was fully aware of this result, having worked in Göttingen during the period when it was discovered (see Chapter 10 below).

The above analysis may be repeated for an implicit discretisation of the form

$$\frac{Q_m^{n+1} - Q_m^n}{\Delta t} + \frac{c}{2}\left(\frac{Q_{m+1}^n - Q_{m-1}^n}{2\Delta x} + \frac{Q_{m+1}^{n+1} - Q_{m-1}^{n+1}}{2\Delta x}\right) = 0. \tag{5.14}$$

This is called the six-point Crank–Nicholson scheme (Kalnay, 2003). Then the phase speed C of the numerical solution is given by

$$C = \frac{2}{k\Delta x} \tan^{-1} \left[\left(\frac{c\Delta t}{2\Delta x} \right) \sin k\Delta x \right]. \tag{5.15}$$

Since this equation contains an inverse tangent term instead of the inverse sine occurring in (5.12), the numerical phase speed C is always real, so the scheme is *unconditionally stable*. It is easily shown that $C \le c$ and that $C \to \pi/k\Delta t$ as $c \to \infty$. Thus, the implicit scheme slows down the faster waves.

The strongest constraint imposed by (5.13) is for the highest wave-speed c occurring in the system. For the atmosphere, the speed of external gravity waves may be estimated as

$$c = \bar{u} + \sqrt{gH}$$

where $\bar{u}$ is the advecting speed and H is the scale height. We may assume $\bar{u} < 100\,\mathrm{m\,s^{-1}}$ and $\sqrt{gH} < 300\,\mathrm{m\,s^{-1}}$, so a safe maximum value for c is $400\,\mathrm{m\,s^{-1}}$. Then the maximum allowable time step for various spatial grid sizes may be estimated:

Δx:	200 km	100 km	20 km	10 km
Δt:	500 s	250 s	50 s	25 s

In the two-dimensional case, the stability criterion is more stringent: we need to choose a time step that is $\sqrt{2}$ times smaller than that permitted in the one-dimensional case (Mesinger and Arakawa, 1976, p. 35). Thus, for Richardson's forecast, which had $\Delta x \approx 200\,\mathrm{km}$, the maximum permissible time step would be 350 s or about six minutes. The calculations of Richardson were confined to the evaluation of the initial tendencies (at 0700 UTC on 20 May 1910). He multiplied these by a time interval $2\Delta t = 6\,\mathrm{h}$ to represent the change over the six-hour period centred at 0700 UTC. In modern terminology, the time-step is specified as the interval between adjacent evaluations of the variables; thus, the time-step used by Richardson was three hours, not six hours as is so often stated. A three-hour step was also chosen by him in describing his fantastic forecast factory (*WPNP*, p. 219). Clearly, Richardson's choice of time step failed utterly to satisfy the CFL stability criterion. However, this was not the reason for his unrealistic initial tendencies.

5.4 The Richardson grid

We described the discretisation of the horizontal spatial derivatives in the previous chapter. The Richardson grid (Fig. 4.2) was specially constructed so that the approximations were centred, ensuring second-order accuracy. The atmosphere is divided into a small number of discrete horizontal layers (depicted in Fig. 5.2 below). Each horizontal layer is partitioned into rectangular boxes or grid-cells. The sides of each box are bounded by parallels of latitude and meridians of longitude. We denote the increment between adjacent boxes in the east–west direction by $\Delta\lambda$ and in the north–south direction by $\Delta\phi$. Richardson selected boxes with sides of length 200 km from south to north; this value corresponds to $1.8°$. The grid-size chosen depends on the range of meteorological phenomena being studied. Richardson was guided in his choice by the density of existing observing stations.

It is desirable for the boxes to be nearly square; to allow for convergence of the meridians towards the poles, Richardson proposed choosing the increment in longitude so that the number of boxes circling the globe was a high power of two, allowing repeated subdivision of the number as the poles were approached. He suggested using 128 boxes (64 with pressure and 64 with momentum), corresponding to $\Delta\lambda = 2°48'45''$. The frontispiece of *WPNP* shows a map of Europe divided into cells in accordance with this scheme (see Fig. 1.6 on page 21). The first omission of alternate cells should be made at about $63°$N; curiously, the suggested reduction is not depicted on Richardson's map. In the example forecast which forms the focal point of the book, he chose an increment in longitude $\Delta\lambda = 3°$. This choice yields grid-boxes that are roughly square in the region of interest (around $50°$ N). Richardson considered this to be an inferior choice, remarking plaintively that the use of a power of two 'was not thought of until after the $3°$ difference of longitude had been used in the example of Chapter 9' (*WPNP*, p. 19). However, he did choose $\Delta\lambda = (360/128) = 2.8125°$ for his barotropic forecast in Chapter 2 of *WPNP*.

Richardson considered the possibility of using a time-staggered grid, with different variables specified at different times. He concluded that the disadvantages outweighed the advantages and decided to tabulate all quantities at the same set of times. In fact, the Richardson grid splits neatly into two sub-grids, one with pressures at even times ($t = 2n\Delta t$) and momenta at odd times and a conjugate one with the remaining quantities. The subscripts 'O' and 'E' indicate odd and even time-steps respectively (see Fig. 5.1 below). We could envisage an integration proceeding on one of the sub-grids, completely independently of the values on the other one. Time-staggered grids were investigated much later by Eliassen (Mesinger and Arakawa, 1976, §4.4). Of course, this ideal separation is valid only for the linear Laplace tidal equations. For more realistic equations, the additional terms introduce inter-dependencies between the sub-grids.

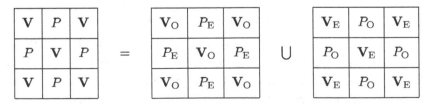

Figure 5.1 Splitting of the grid into two sub-grids for odd and even time-steps.

5.5 The equations for the strata

Richardson chose to divide the atmosphere into five strata of approximately equal mass, separated by horizontal surfaces at 2.0, 4.2, 7.2 and 11.8 km, corresponding to the mean heights (over Europe) of the 800, 600, 400 and 200 hPa surfaces. He discusses this choice in *WPNP*, §3/2. It is desirable to have a surface near the tropopause, one stratum for the planetary boundary layer and at least two more for the troposphere above the boundary layer. The requirement for two layers in the troposphere was 'to represent the convergence of currents at the bottom of a cyclone and the divergence at the top' (*WPNP*, p. 16). Thus, Richardson explicitly recognised the importance of the Dines mechanism. Taking layers of equal mass simplified the treatment of processes like radiation, and the particular choice of surfaces at approximately 200, 400, 600 and 800 hPa greatly facilitated the extraction of initial data from the charts and tables of Bjerknes. The strata are depicted in Fig. 5.2. Each horizontal layer was divided up into rectangular boxes or grid-cells of size $\Delta\lambda = 3°$ and $\Delta\phi = 1.8°$.

The fixed heights of the horizontal interfaces between Richardson's 'conventional strata', which correspond approximately to pressures of 200, 400, 600 and 800 hPa, will be denoted respectively by z_1, z_2, z_3 and z_4, all constants. The variable height of the Earth's surface will be written $z_5 = h(\lambda, \phi)$. Variables at these five levels will be denoted by corresponding indices 1–5. Where convenient, values at the surface of the Earth may be indicated by subscript S. Thus, p_S is the surface pressure, i.e. the pressure at $z = h$ (Richardson uses p_G).

Following Richardson, the equations of motion are now integrated with respect to height across each stratum, to obtain expressions applying to the stratum as a whole. The continuous vertical variation is thus averaged out and equations for the mean values in each of the five layers are derived. Quantities derived by vertical integration in this way will be denoted by capitals:

$$R = \int \rho \, dz \qquad P = \int p \, dz \qquad U = \int \rho u \, dz \qquad V = \int \rho v \, dz.$$

The stratum is specified by the index corresponding to the *lower* level; thus, for example,

$$R_3 = \int_{z_3}^{z_2} \rho \, dz. \qquad (5.16)$$

In calculating horizontal derivatives of mean values for the lowest layer, allowance must be made for the variation of the height h of the Earth's surface. For the other layers, the limits are independent of x and y. Richardson specified the pressure at the level interfaces (Fig. 5.2) to fit with the hydrostatic equation. Substituting this

The solution algorithm

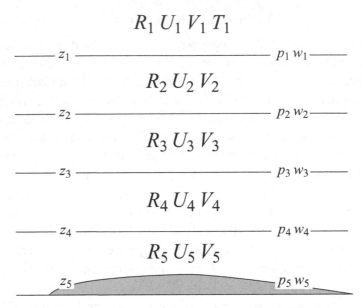

Figure 5.2 Vertical stratification: R, U and V are specified for each layer, p and w are evaluated at the interfaces. The stratospheric temperature T_1 is also required.

equation in (5.16), we have

$$R_3 = -\int_{z_3}^{z_2} \frac{1}{g} \frac{\partial p}{\partial z} \, dz = \frac{p_3 - p_2}{g}.$$

This arrangement in the vertical is similar to the Charney–Phillips grid, which is now more popular than the Lorenz grid (Kalnay, 2003). The vertical velocity w is also specified at level interfaces.

The continuity equation (2.2) will now be integrated in the vertical. Taking, for example, the stratum bounded below and above by z_B and z_T and denoting these limits by B and T, we get

$$\frac{\partial}{\partial t} \int_{B}^{T} \rho \, dz + \frac{\partial}{\partial x} \int_{B}^{T} \rho u \, dz + \frac{\partial}{\partial y} \int_{B}^{T} \rho v \, dz - \frac{\tan \phi}{a} \int_{B}^{T} \rho v \, dz + [\rho w]_T - [\rho w]_B = 0.$$

Using the definitions of R, U and V this equation may be written

$$\frac{\partial R}{\partial t} + \frac{\partial U}{\partial x} + \frac{\partial V}{\partial y} - \frac{V \tan \phi}{a} + [\rho w]_T - [\rho w]_B = 0. \qquad (5.17)$$

The equations for all the upper layers are of similar form. For the lowest layer the slope of the bottom boundary must be considered, and the term $-[\rho w]_S$ is cancelled by a term $[\rho \mathbf{V} \cdot \nabla h]_S$.

The vertical integration of the horizontal equations of motion is performed in the same manner. The equations are

$$\frac{\partial \rho u}{\partial t} + \frac{\partial \rho u^2}{\partial x} + \frac{\partial \rho uv}{\partial y} + \frac{\partial \rho uw}{\partial z} - \left(f + \frac{2u \tan \phi}{a} \right) \rho v + \frac{\partial p}{\partial x} = 0$$

$$\frac{\partial \rho v}{\partial t} + \frac{\partial \rho vu}{\partial x} + \frac{\partial \rho v^2}{\partial y} + \frac{\partial \rho vw}{\partial z} + f\rho u + \frac{(\rho u^2 - \rho v^2)\tan \phi}{a} + \frac{\partial p}{\partial y} = 0.$$

To express the result in terms of the variables R, U and V it is necessary to make an approximation in the horizontal flux terms. For example, in the zonal momentum equation, the terms

$$\int_B^T \frac{\partial}{\partial x} \rho u^2 \, dz + \int_B^T \frac{\partial}{\partial y} \rho uv \, dz$$

occur. If the velocity is constant throughout the stratum then $u = U/R$ and $v = V/R$ and these terms become

$$\frac{\partial}{\partial x}\left(\frac{U^2}{R} \right) + \frac{\partial}{\partial y}\left(\frac{UV}{R} \right).$$

With this approximation the equations for the stratum (z_B, z_T) are

$$\frac{\partial U}{\partial t} + \frac{\partial}{\partial x}\left(\frac{U^2}{R} \right) + \frac{\partial}{\partial y}\left(\frac{UV}{R} \right) + [\rho uw]_T - [\rho uw]_B - fV + \frac{2UV \tan \phi}{aR} + \frac{\partial P}{\partial x} = 0$$

$$\tag{5.18}$$

$$\frac{\partial V}{\partial t} + \frac{\partial}{\partial x}\left(\frac{UV}{R} \right) + \frac{\partial}{\partial y}\left(\frac{V^2}{R} \right) + [\rho vw]_T - [\rho vw]_B + fU + \frac{(U^2 - V^2)\tan \phi}{aR} + \frac{\partial P}{\partial y} = 0.$$

$$\tag{5.19}$$

The equations for all the upper layers are of similar form. For the lowest layer the slope of the bottom boundary must be considered, and the equations become

$$\frac{\partial U}{\partial t} + \frac{\partial}{\partial x}\left(\frac{U^2}{R} \right) + \frac{\partial}{\partial y}\left(\frac{UV}{R} \right) + [\rho uw]_4 - fV + \frac{2UV \tan \phi}{aR} + \frac{\partial P}{\partial x} + \left[p\frac{\partial h}{\partial x} \right]_S = 0$$

$$\frac{\partial V}{\partial t} + \frac{\partial}{\partial x}\left(\frac{UV}{R} \right) + \frac{\partial}{\partial y}\left(\frac{V^2}{R} \right) + [\rho vw]_4 + fU + \frac{(U^2 - V^2)\tan \phi}{aR} + \frac{\partial P}{\partial y} + \left[p\frac{\partial h}{\partial y} \right]_S = 0.$$

These equations correspond to (11) and (12) on page 34 of *WPNP*.

The gradient with height of vertical velocity is given by (2.20) which, for convenience, we repeat here:

$$\frac{\partial w}{\partial z} = -\nabla \cdot \mathbf{V} + \frac{1}{\gamma p} \int_0^p \left(\nabla \cdot \mathbf{V} - \frac{\partial \mathbf{V}}{\partial p} \cdot \nabla p \right) dp. \tag{5.20}$$

This corresponds to (9) on page 124 of *WPNP*. The quantity $\partial w/\partial z$ is computed at the centre of each stratum. In the stratosphere, it is assumed to be constant with height. The stratospheric temperature is predicted using the simple equation

$$\frac{\partial T_1}{\partial t} = T_1 \left(\frac{\partial w}{\partial z} \right)_1 , \tag{5.21}$$

which was derived in §2.4.

The numerical integration of the equations is carried out by a step-by-step procedure – an algorithm – which produces later values from earlier ones. Richardson took pains to devise a 'lattice-reproducing' scheme, which ensured that where a particular variable was given at an initial time, the corresponding value at a later time *at the same point* could be computed.

5.6 The computational algorithm

In Chapter 8 of *WPNP*, Richardson sets down the step-by-step sequence of calculations necessary to carry his forecast forward in time. This chapter is 25 pages long, and includes a number of digressions and detailed discussion of hydrological and thermodynamic processes which have minimal impact on the forecast. Such abundance of minor detail makes the main thread of his procedure hard to follow; we will attempt a more succinct description in this section.

Let us assume that all the dependent variables are known at time $t = n\Delta t$ and also at the previous time level $t = (n-1)\Delta t$. The advancement to the next time level, $t = (n+1)\Delta t$, requires both prognostic and diagnostic components (these terms, borrowed from medicine, were introduced by Vilhelm Bjerknes). The prognostic variables are (at P-points) the mass per unit area R in each stratum and the stratospheric temperature T_1, and (at M-points) the components U and V of momentum in each stratum. Once these quantities are known for a particular moment, all the auxilliary fields (temperature, divergence, vertical velocity, etc.) for that moment can be calculated from diagnostic relationships.

$\star$ $\qquad$ $\star$ $\qquad$ $\star$ $\qquad$ $\star$ $\qquad$ $\star$

The time-stepping calculations are done in a large loop that is repeated as often as required to reach the forecast span: for a span t_{max} and time-step Δt we need $N = t_{max}/\Delta t$ steps. The sequence of calculations will now be given. For each step, the number of the relevant computing form in *WPNP* is indicated [in brackets]. First we consider the P-points.

(i) [P_I] The layer integral of pressure is calculated:

$$P = \Delta z \cdot \Delta p / \Delta \log p$$

Here Δ represents the difference in value across the layer. For the top layer $P = \Re T_1/g$. The density integral is also calculated from

$$R = \frac{\Delta p}{g}.$$

(ii) [P$_I$] Mean values of various quantities are calculated for each stratum except the uppermost one; e.g.,

$$\bar{p} = \frac{P}{\Delta z} \qquad \bar{\rho} = \frac{R}{\Delta z} \qquad \bar{T} = \frac{\bar{p}}{\Re \bar{\rho}}.$$

We also require $\tilde{p}_1 = e^{-1} p_1$, the pressure at $z = z_1 + \Re T_1/g$, one scale-height above the tropopause.

(iii) [P$_{XIII}$] The divergence of momentum $\mathbf{U} = (U, V)$ is computed for each level. Because of the grid staggering, values of $\nabla \cdot \mathbf{U}$ are obtained at the P-points.

(iv) [P$_{XIII}$] The values of $\nabla \cdot \mathbf{U}$ in the column above each P-point are summed up and the total multiplied by $-g$ to give the surface pressure tendency

$$\frac{\partial p_S}{\partial t} = -g \sum_{\text{all}}^{\text{strata}} \nabla \cdot \mathbf{U}.$$

Calculation of the interface pressure tendencies must await the availability of the vertical velocity.

(v) [P$_{XV}$] The divergence of velocity $\mathbf{V} = (u, v)$, denoted $\delta = \nabla \cdot \mathbf{V}$, is calculated using the following approximations for mean velocity in each layer:

$$u = U/R \qquad v = V/R.$$

The velocity divergence is used in the calculation of the vertical velocity. The quantities N_1 and N_2, which are measures of noise, are computed at this stage.

(vi) [P$_{XVI}$] The vertical velocity gradient $\partial w/\partial z$ in each layer is now calculated using (5.20). The vertical velocity at the surface is determined from

$$w_S = \mathbf{v}_S \cdot \nabla h$$

where we assume $\mathbf{v}_S = k \mathbf{V}_5/R_5$ with $k = 0.2$. Then it is a straightforward matter to calculate w at each interface, working upward from the bottom.

(vii) [P$_{XIV}$] The tendency of the stratospheric temperature T_1 is calculated next, using (5.21)

$$\frac{\partial T_1}{\partial t} = T_1 \left(\frac{\partial w}{\partial z} \right)_1.$$

(viii) [P$_{XVI}$] The temperature at each interface is calculated by linear interpolation. Then the density there is computed using the gas law, after which the vertical momentum ρw at each interface can be obtained. This is required for vertical mass flux in the continuity equation.

(ix) [P$_{XIII}$] The tendency of the density integral R is now obtained using the continuity equation

$$\frac{\partial R}{\partial t} + \left(\frac{\partial U}{\partial x} + \frac{\partial V}{\partial y} - \frac{V \tan \phi}{a} \right) + [\rho w]_{\mathrm{T}} - [\rho w]_{\mathrm{B}} = 0.$$

(x) [P$_{XIII}$] The final calculation at P-points is the tendency of pressure at each interface, obtained from

$$\left(\frac{\partial p}{\partial t} \right)_K = g \sum_{k=1}^{K} \left(\frac{\partial R_k}{\partial t} \right).$$

The surface pressure tendency, already computed in step (iv), is confirmed here.

This completes the calculations required at the P-points.

We now list the operations at the M-points [*WPNP*, Computing Forms M$_{III}$ and M$_{IV}$]

(i) The pressure gradient is evaluated by calculating the spatial derivatives of the integrated pressure P. As a result of the staggering of the grid, the gradient is obtained at the M-points where it is required. For the lowest stratum there is an extra term due to orography; to reduce numerical errors, this is calculated in combination with the pressure gradient. The x-component is given by

$$\frac{1}{2\Delta x} \left\{ \delta P + \frac{\delta h \cdot \delta p_S}{\delta \log p_S} \right\} \tag{5.22}$$

where δ here represents the difference across a distance $2\Delta x$ (*WPNP*, p. 35). The y-component of the pressure gradient is calculated in an analogous manner.

(ii) The Coriolis terms and the terms in (5.18) and (5.19) involving $\tan \phi$ are evaluated. All the necessary quantities are available at the relevant points.

(iii) The horizontal flux terms in (5.18) and (5.19) are calculated. It is necessary to approximate the derivatives by differences over a distance $4\Delta x$ or $4\Delta y$ in computing these terms.

(iv) The vertical flux terms are calculated. The vertical velocity at M-points is obtained by interpolation from surrounding P-points; the horizontal velocities at the interfaces are evaluated by averaging the values in the layers above and below. The momentum flux above the uppermost layer is assumed to vanish.

(v) The tendencies of momenta, $\partial U / \partial t$ and $\partial V / \partial t$ may now be calculated, as all the other terms in (5.18) and (5.19) are available.

The tendencies of all prognostic variables are now known, and it is possible to update all the fields to the time $(n + 1)\Delta t$. When this is done, the entire sequence of operations may be repeated in another time-step.

★   ★   ★   ★   ★

The above procedure is, in Richardson's term, lattice-reproducing. In principle, the sequence of operations may be repeated indefinitely, to compute a forecast of arbitrary length. However, several difficulties arise in practice. When the domain is of limited geographical extent, special treatment of the lateral boundaries is required (see p. 229 below). In the example presented in Chapter 7, we hold the values at the boundaries constant at their initial values. Richardson discussed lateral boundaries in Ch. VII of *WPNP*, but did not encounter this issue in his forecast, which was confined to the calculation of the initial tendencies.

A further problem of great practical significance is the generation of numerical noise during the integration. To control this, various smoothing operations are applied to the values calculated at each time step. We will describe the smoothing in more detail in Chapter 9.

6

Observations and initial fields

*... the icy layers of the upper atmosphere contain conundrums enough to
be worthy of humanity's greatest efforts ...* (*Hergesell, 1905*)

We have seen in the last chapter how Richardson constructed a complete algorithm
for integrating the equations of motion and, in the following one, we will study
the application of his method to an actual weather situation. To do that, initial
conditions are required, and we describe in this chapter the means of obtaining
them. We outline the emergence of aerology, the study of the upper atmosphere.
We describe an ingenious instrument for making aerological observations. Then
we consider some results of Bjerknes' 'diagnostic programme'. Finally, we discuss
the preparation of the initial fields for the numerical forecast.

6.1 Aerological observations

During the nineteenth century, measurements of atmospheric conditions at the
Earth's surface were made on a regular basis, and daily surface weather maps
were issued by several meteorological institutes. Observations at elevated locations
were much more difficult to obtain. A number of mountain stations were established
towards the end of the century: Mt. Washington in 1870, Pike's Peak in 1873, Pic du
Midi in 1886, Fujiyama in 1898 and Zugspitze in 1900 (Khrgian, 1959). However,
climatological conditions at mountain stations have particular characteristics. What
was desirable was an investigation of conditions in the free atmosphere. Several
intrepid balloonists ascended with instruments to measure the vertical structure of
the atmosphere. James Glaisher, of the Royal Observatory in Greenwich, made
numerous ascents in the 1860s to measure the variations of temperature and humidity with height. Such explorations were both expensive and dangerous; indeed,
there were a number of fatalities. The credit for initiating systematic observations
of the upper air goes to Lawrence Rotch of Blue Hill Observatory, Boston, who,

97

around 1895, started the method of attaching self-recording instruments to kites. Kites and tethered balloons had a limited vertical range, and the steel piano-wire used to control them occasionally snapped, with serious consequences.[1]

Sounding balloons, which carried self-registering instruments aloft as they rose by natural buoyancy, were proposed by George Besançon and Gustave Hermite (a nephew of the renowned mathematician, Charles Hermite) as a powerful, practical technique for exploring the higher reaches of the atmosphere. These balloons were tracked by two theodolites, separated by a baseline of 1 km or more. Balloons launched at night were equipped with a small light to render them visible. The wind speed and direction at each level could be deduced from the azimuth and elevation bearings of the theodolites. A large specially-designed slide-rule was used for the calculations, and the winds were available in near 'real-time'. Pilot balloons were small balloons without any instruments attached. They were named for the aviators who launched them prior to take-off to see how the winds aloft were blowing. For the temperature and humidity, the recording instrument had to be recovered. This normally took at least several days. Such data were clearly of no relevance for operational weather prediction, but would prove hugely beneficial to researchers.

In addition to the government-funded activities, several aerological observatories were established by enthusiasts. Thus, the observatories at Blue Hill, Boston, and at Trappes, Paris, were run by Lawrence Rotch and Teisserenc de Bort respectively, using their own financial resources. Aeronautical pioneers also contributed to the advancement of this emerging science.[2] The first instrumented balloon ascent was carried out by Gustave Hermite on 17 September 1892, using a waxed paper balloon. Constant volume balloons halt their ascent when they reach their level of neutral buoyancy. Around 1900, Richard Assmann, Director of the Royal Prussian Observatory in Berlin, revolutionised the sounding technique when he introduced rubber balloons. These expand with height, maintaining an approximately constant rate of ascent. Assuming this rate to be known, the position can be determined by means of a single theodolite. These expanding balloons never reach a position of equilibrium, but continue to rise until they burst. They enable measurements to be made at heights up to 15 or 20 km. They are reliable and relatively inexpensive and are still in use today. The method of upper air observation using sounding balloons came into widespread use from this time. It was now possible to launch balloons simultaneously from several locations, raising the possibility of synoptic aerology (Nebeker, 1995).

---

[1] In one notorious incident an array of kites, launched by Teisserenc de Bort, broke loose and drifted across Paris trailing 7 km of wire. They stopped a steamer and a train and disrupted telegraphic communications with Rennes on the day when the results of the Dreyfus court-martial in that city were anxiously awaited (Ohring, 1964).

[2] Bjerknes (1910) summarised the symbiosis between aviation and aerology: 'The development of aeronautics will make these [aerological] observations not only possible, but also necessary'.

INTERNATIONAL METEOROLOGICAL CONFERENCE, PARIS, 1896

Snellen	Riggenbach	Kesslitz	Biese	Moureaux	Rotch	Fines	Jaubert	Thévenet	Lancaster	Chauveau	Page
Mathias	de Fonvielle	Anguiano			Paulsen			
van Rijckevorsel	Ellis	Hergesell	A. Schmidt	Watzoff		Mohn	Angot	Rykatcheff	Hepites	Billwiller	Erk
Teisserenc de Bort	Hildebrandsson	Scott		Tacchini		Mascart	von Bezold		Rücker

Figure 6.1 International Meteorological Conference, Paris, 1896. (From Shaw, 1932)

At the Conference of the International Meteorological Organization in Paris in 1896 (see Fig. 6.1), an International Commission for Scientific Aeronautics (ICSA) was established. Hugo Hergesell, Director of the Meteorological Institute of Strasbourg, was appointed President of the Commission. Hergesell was actively involved in upper air observations. He was also a consummate diplomat, and succeeded in overcoming the rivalry between the French and German scientists and establishing excellent international collaboration. The tasks of the Commission were to co-ordinate and regulate upper air research in Europe, to establish standards and to organise simultaneous observations of the free atmosphere on 'international aerological days'. Hergesell published a series of reports of the ICSA Conferences. He also published, between 1901 and 1912, some 22 volumes of data acquired during the aerological days. The first such experiment was on 13/14 November 1896, with remarkable results, which Shaw would, much later, refer to as 'the most surprising discovery in the whole history of meteorology' (Shaw, 1932, p. 225). The sonde launched from Paris showed an isothermal layer above 12 km. This was the first indication of the stratosphere, but the measurements were discounted as erroneous and it took several years before the existence of the tropopause and stratosphere were firmly established, independently by Teisserenc de Bort and Richard Assmann. For an interesting account of the events surrounding the discovery of the tropopause, see Hoinka, 1997.

Upper air observations were made only intermittently, typically for one or a few days each month, as agreed by the countries participating in the work of the ICSA. European aerological stations active at this time included Aachen, Bath, Berlin, Copenhagen, De Bilt, Guadalajara, Hamburg, Kontcheiv, Lindenberg, Milan, Munich, Pavia, Pavlovsk, Strasbourg, Trappes, Uccle, Vienna and Zurich. The international aerological days, or 'balloon days', were normally on the first Thursday of each month. In three months of each year the adjacent Wednesday and Friday were added, giving three consecutive days and, once a year, six consecutive days, the 'international week'. The co-operation came to a sudden end with the First World War. Although the radiosonde was invented in 1927, it was not until after the Second World War that a real synoptic upper air network was established in Europe.

6.2 Dines' meteorograph

William Henry Dines (Fig. 6.2) was a master in the design and construction of meteorological instruments. He was active on the Wind-Force Committee that was set up following the Tay Bridge disaster. A train fell into the river when the bridge was blown down in a storm in December 1879, with the loss of over 100 lives. This provided a strong incentive for the development of instruments capable of measuring the wind accurately. One result was Dines' pressure-tube anemometer, an ingenious construction that gives a continuous record of the wind speed and direction, with detailed reading of gusts and lulls. Dines also expended considerable energy investigating solar and terrestrial radiation, and undertook some more general studies of atmospheric structure. However, his main contributions were to the study of the upper atmosphere. Although it was not until 1908 that regular aerological observations began in England, Dines began his investigatons of observing techniques some years earlier. Shortly after Sir Napier Shaw took charge of the Met Office in 1900, he encouraged Dines to undertake observational studies of the upper air. Dines had exceptional flair in designing meteorological instruments, and his meteorograph, which we will describe now, was a masterpiece of economy and efficiency (Dines, 1909).

Registering balloons measured pressure, temperature and humidity during their ascent and recorded the values by means of an instrument called a meteorograph. Analysis of the values was dependent upon this instrument being found after its descent by parachute. To encourage finders to return the device, a notice was attached to the instrument, offering a reward (see Fig. 6.3). According to Dines (1919) 'it is astonishing how many are returned; the [European] continental stations do not lose more than one out of ten, but in England many fall in the sea and the loss reaches 30 or 40 per cent.'

Figure 6.2 W. H. Dines (1855–1927). From *Collected Scientific Papers of William Henry Dines*, Royal Meteorological Society, 1931.

Wind speeds and directions at various heights were deduced by following the course of the ascending balloon. Obviously, this became impossible once the balloon entered cloud. But these observations were available promptly, whereas the pressure and temperature data were obtained only when the instrument was found and returned after its descent. As Dines wrote (1919), 'The mere sending up of a registering instrument attached to a balloon does not necessarily mean a good observation. The instrument may never be found, the clock may stop, the pen may not write, the finder may efface the record; there are many possibilities of failure.' On 20 May 1910, for example, the observation for Vienna comprised only winds; no pressures or temperatures were available, the registering balloon being recorded as *bis heute noch nicht gefunden* (Hergesell, 1913). It is unlikely to be found now.

The instruments were not standardised, different systems being used in different countries. In the French and German instruments, records were made on smoked

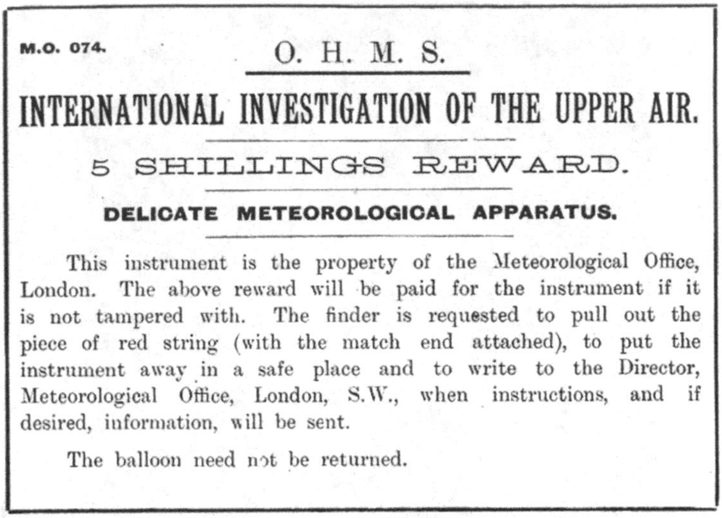

M.O. 074.   O. H. M. S.

INTERNATIONAL INVESTIGATION OF THE UPPER AIR.

5 SHILLINGS REWARD.

DELICATE METEOROLOGICAL APPARATUS.

This instrument is the property of the Meteorological Office, London. The above reward will be paid for the instrument if it is not tampered with. The finder is requested to pull out the piece of red string (with the match end attached), to put the instrument away in a safe place and to write to the Director, Meteorological Office, London, S.W., when instructions, and if desired, information, will be sent.

The balloon need not be returned.

Figure 6.3 Label attached to meteorographs of the Met Office, offering a five shilling reward for return of the instrument. (From Dines, 1912)

paper secured to a drum which was turned by a clock. The instruments weighed about 1 kg and required a ballon of diameter 2 m to carry them. The meteorograph designed by Dines was of elegant simplicity, inexpensive to make and weighing about an ounce (30 g). Its cost was only £1, in comparison to £15 or £20 for instruments used in other countries. It carried no clock but recorded temperature as a function of pressure. In favourable circumstances, the instrument could be carried by a small balloon to a height of 20 km.

A diagram of Dines' meteorograph is presented in Fig. 6.4. The frame is cut from a single piece of metal, the end B being turned down at right angles to allow it to open and close like a pair of scissors as the aneroid box A expands or contracts with changing pressure. One side of the frame carries two steel points (pens) E and L and on the other there is a small square metal plate the size of a postage stamp on which they etch marks as they move. Pen L, attached to bar H records the pressure. The lever DCF carrying pen E is free to pivot about C; as the temperature falls, the strip M of German silver contracts, so point D moves downward and pen E upward. With uniform temperature, a decrease of pressure causes two parallel scratches on the plate, whereas a change of temperature causes a change in the distance between the scratches. After retrieval, the small plate is removed from the instrument and examined under a microscope. The distance between the marks indicates the temperature as a function of the pressure. A third element to measure humidity can easily be added. It is similar to the temperature element but the expanding metal strip is replaced by strands of hair, the length of which is sensitive

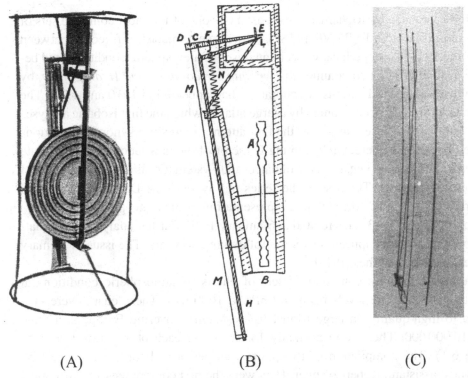

Figure 6.4 Dines' meteorograph for upper air soundings. (A) Photograph of instrument showing the aneroid box. (B) Schematic diagram indicating the operating principle. (C) Meteorogram, showing scratches on a small metal plate. (Dines, 1909, 1912)

to the relative humidity of the air. The right-hand panel in Fig. 6.4 is a meteorogram produced by Dines' instrument.

The height may be obtained by means of the hydrostatic relationship so that, for a column between pressures p_1 and p_2, the thickness is

$$\Delta z = z_2 - z_1 = \frac{\Re \bar{T}}{g} (\log p_1 - \log p_2)$$

where $\bar{T}$ is the mean temperature of the column. This can easily be evaluated using special graph-paper. A detailed example was given in Dines (1909). He also gave an estimate of the accuracy of his meteorograph: the error in temperature is generally less than 3°C; the pressure error less than 10 mm Hg (about 13 hPa). This means that at 10 km the height error may be up to 400 m, and at 20 km up to about 1500 m. These are quite large errors and cause considerable uncertainty, particularly for the analysis at higher levels.

6.3 The Leipzig charts

The forecast made by Richardson was based on 'one of the most complete sets of observations on record' (*WPNP*, p. 181). At the time he made this forecast (between 1916 and 1918) a comprehensive set of analyses of atmospheric conditions had become available. The two volumes of *Dynamic Meteorology and Hydrography*, by Vilhelm Bjerknes and various collaborators, had appeared in 1910 and 1911. The second volume was accompanied by a large atlas in which the first isobaric analyses were published. These maps were the first attempt to analyse synoptic conditions in the upper atmosphere. On becoming Director of the new Geophysical Institute in Leipzig, Bjerknes began a consolidated and systematic diagnostic analysis of the aerological data. The first of the series of 'Synoptische Darstellungen atmosphärischer Zustände' (the synoptic representation of the atmospheric conditions) was published in 1913. This related to 6 January 1910. Further analyses, also relating to the year 1910, appeared over the following two years. The issue of primary interest to us is Bjerknes, 1914b.

Bjerknes' analyses consisted of sets of charts of atmospheric conditions at ten standard pressure levels from 100 hPa to 1000 hPa. These charts were produced to high-quality, in large format (64 × 40 cm), covering Europe at a scale of 1:10 000 000. There were normally 14 charts for each observation time (see Table 6.1). The compilation of the charts was performed for the most part by Bjerknes' assistant, Robert Wenger. They were the first comprehensive aerological analyses ever published. They enabled Bjerknes to study the three-dimensional evolution of atmospheric conditions, and to test his prognostic methods that were based on graphical techniques (Bjerknes *et al.*, 1910, 1911). He was convinced that, ultimately, charts such as these would be the basis for a rational forecasting scheme. In this he was correct, although perhaps not in the manner he envisaged: the charts provided Richardson with the data required for his arithmetical forecasting procedure.

The 'international aerological days' were normally on the first Thursday of each month. In normal circumstances, there would have been a balloon day on Thursday 5 May 1910. It is interesting that the observational period for May 1910 was postponed to coincide with the passage of Halley's comet. There was some speculation that the comet might cause a detectable response in the atmospheric conditions, and what would now be called an 'intensive observing period' was undertaken. For example, on 19 May, a series of hourly ascents over a period of 24 hours was carried out at Manchester to ascertain the diurnal variation of temperature. The comet passed between the Earth and the Sun on 18 May; as the tail curved slightly backwards, the passage of the Earth through it occurred a little later, on the 20th, the day chosen by Richardson (Lancaster-Browne, 1985). Comets are popularly thought to portend dramatic events; one may say that on this occasion a comet was associated with an

Table 6.1 *Analysed charts in* Synoptische Darstellungen atmosphärischer Zustände. *Jahrgang 1910, Heft 3 (Bjerknes et al., 1914b).*

The Roman numerals in column 2 are the level indicators used by Bjerknes. For 0700, 20 May 1910 the 100 mb analysis is missing, due to lack of sufficient observational data at this level.

Chart	Level	Content
1		Sea level pressure (mm Hg) and temperature (°C)
2		Surface streamlines and isotachs (m/s)
3		Cloud cover and precipitation
4	X	1000 mb height and 1000–900 relative topography
5	IX	900 mb height and 900–800 relative topography
6	VIII	800 mb height and 800–700 relative topography
7	VII	700 mb height and 700–600 relative topography
8	VI	600 mb height and 600–500 relative topography
9	V	500 mb height and 500–400 relative topography
10	IV	400 mb height and 400–300 relative topography
11	III	300 mb height and 300–200 relative topography
12	II	200 mb height and 200–100 relative topography
13	I	100 mb height
14		Tropopause height

event of great significance for meteorology, though not due to its having any direct influence on the atmosphere.

The date and time chosen by Richardson for his initial data was 20 May 1910, 0700 UTC. During the three-day observing period there ascended altogether 73 registering balloons (33 of which included wind observations), 35 kite and captive balloons, 81 pilot balloons and 4 manned balloons. Aerological observations were reported for the following locations: Aachen, Bergen, Christiania, Copenhagen, De Bilt, Ekaterinburg, Friedrichshafen, Hamburg, HMS Dinara (near Pola), Lindenberg, Manchester, Munich, Nizhni-Olchedaev, Omsk, Pavia, Pavlovsk, Petersfield, Puy de Dome, Pyrton Hill, Strasbourg, Stuttgart, Tenerife, Trappes, Uccle, Vienna, Vigne di Valle, Zurich and, from outside Europe, Apia, Blue Hill and Mt. Weather. The stations of most relevance for Richardson's forecast are indicated in Fig. 6.9 on page 110 below. The soundings and reports of upper level winds over western Europe for 0700 on 20 May 1910 are given in Table 6.2. The full compilation of observations occupies more than one hundred pages in Hergesell (1913). The observations are tabulated in a compressed form in Bjerknes, 1914b.

The weather conditions during the period 18–20 May 1910 were summarised in Hergesell's publication:

Table 6.2 *Upper Air Observations from registering balloons, pilot balloons and kites for 0700 UTC, 20 May, 1910.*

For the full reports, see Hergesell, 1913.

Location of launch	Minimum pressure (mm Hg)	Maximum height (metres)	Temperature (°C)	Instrument
Aachen	360	6 000	−15.3	Registering balloon
Bergen		9 600		Pilot balloon
Christiania		8 100		Pilot balloon
Copenhagen		12 020		Pilot balloon
Friedrichshaven	460	4 110	−3.9	Kite balloon
Hamburg	195	10 410	−50.1	Registering balloon
Lindenberg	299	7 420	−25.9	Registering balloon
Munich	186	10 600	−55.6	Registering balloon
Nizhni−Olchedaev	632	1 580	5.9	Captive balloon/kite
Pavia	108	13 850	−63.7	Registering balloon
Pavlovsk	132	12 560	−46.5	Registering balloon
Pyrton Hill	69	17 600	−45.5	Registering balloon
Strasbourg	85	15 530	−54.2	Registering balloon
Tenerife		4 710		Pilot balloon
Uccle	91	14 980	−57.8	Registering balloon
Vienna		19 700		Reg. balloon (lost)
Zurich	118	13 450	−47.7	Registering balloon

The distribution of atmospheric pressure was very irregular on the days of the ascents and, consequently, there were frequent thunderstorms, especially in western and central Europe. A cyclone moved northwards from the Bay of Biscay while a weak minimum drifted westwards from the Adriatic, gradually intensifying and an anti-cyclone over Scandanavia gradually increased in strength.   *(Hergesell, 1913)*

The Leipzig publication contains 13 charts for the time in question: sea-level pressure and surface temperature, surface streamlines and isotachs, cloud and precipitation, geopotential heights and thicknesses for nine standard levels at 100 hPa intervals from 1000 hPa to 200 hPa, and tropopause height (see Table 6.1). The sea-level pressure and surface temperature chart is reproduced in Fig. 6.5 and the 500 hPa height in Fig. 6.6.[3]

Data coverage was reasonable at the surface; for the upper levels, the number of observations is seriously limited, leaving great uncertainty over much of the area of interest. In a commentary on the analysis, Wenger wrote that there was no cause to doubt the reliability of any of the ascents. He further commented that 'the good agreement of the wind vectors with the topography of the main isobaric levels' was

---

[3] The full series of charts is available online at http://maths.ucd.ie/~plynch/Dream/Leipzig-Charts.html.

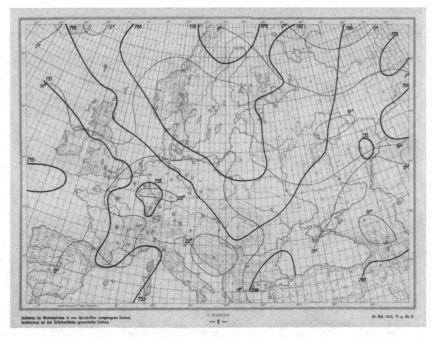

Figure 6.5 Bjerknes' analysis of sea-level pressure (solid lines, mm Hg) and surface temperature (dashed lines, °C) for 0700 UTC on 20 May 1910.

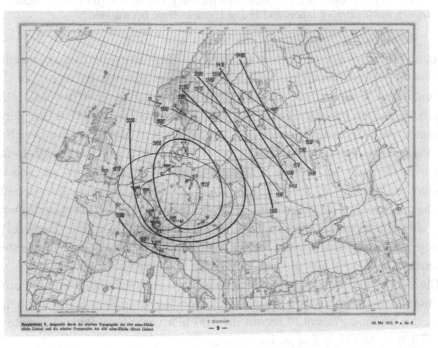

Figure 6.6 Bjerknes' analysis of 500 hPa height (heavy lines, dam) and 500–400 hPa relative topography (light lines, dam) for 0700 UTC on 20 May 1910.

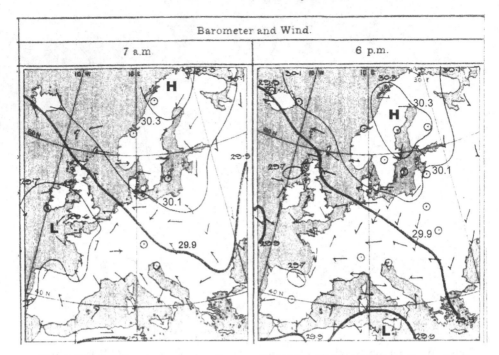

Figure 6.7 Surface analysis over Europe for Friday 20 May 1910. Left: sea-level pressure and surface wind at 0700 UTC. Right: sea-level pressure and surface wind at 1800 UTC. Pressures are in inches of mercury. (From UKMO *Daily Weather Report.* Some contour labels and H and L marks have been added.)

a ground for confidence in the pressure analysis. Thus, he explicitly recognised that the flow in the free atmosphere should be close to geostrophic balance.

Conditions at the surface are also shown in the analysis of the Met Office (Fig. 6.7). The left panel shows the sea level pressure at 0700 UTC. It is in general agreement with Bjerknes' analysis (Fig. 6.5). There is high pressure over Scandanavia and low pressure over Biscay, associated with a generally south-easterly drift over Germany and France. The right panel of Fig. 6.7 shows the sea-level pressure at 1800 UTC on the same day. There is little change in the overall pattern and the pressure over Bavaria remains essentially unchanged. Winds were generally light although a number of thunderstorms were reported. It might reasonably be expected that a pressure forecast for this region would be consistent with the steady barometer. As we will see, this was not true of Richardson's forecast.

The Leipzig publication did not include a chart of 100 hPa height for the time in question. Only one registering balloon, that launched from Pyrton Hill in England, reached a height sufficient to record the 100 hPa value (see Table 6.2; note that 100 hPa ≈ 75 mm Hg). The problem of accurate analysis at this level, with so few observations, is vividly illustrated by Bjerknes' 100 hPa analyses for the previous

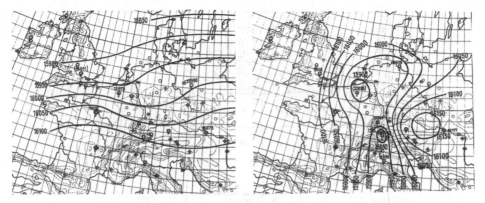

Figure 6.8 Bjerknes' height analyses at 100 hPa at two times on 19 May 1910.
(a) Analysis at 0200 UTC. (b) Analysis at 0700 UTC. Note that only five hours
separate the two analysis times.

day: the charts for 0200 UTC and 0700 UTC are shown in Fig. 6.8; they show
dramatically different flow patterns, and cannot be reconciled with each other considering that their 'valid times' are only five hours apart. The observational errors
may have been due to the 'radiation effect', heating of the thermometer by direct or
reflected sunlight. This was one of the most serious defects of early meteorographic
equipment.

6.4 Preparation of the initial fields

6.4.1 Richardson's analysis

Using the most complete set of observations available to him, Richardson derived
the values of the prognostic parameters at a small number of grid points in central
Europe. The values he obtained were presented in his 'Table of Initial Distribution'
(*WPNP*, p. 185). Richardson chose to divide the atmosphere into five layers, centred
approximately at pressures 900, 700, 500, 300 and 100 hPa (see Fig. 5.2 on page 90).
He divided each layer into boxes and assumed that the value of a variable in each
box could be represented by its value at the central point. The boxes were separated
by $\Delta\lambda = 3.0°$ in longitude and $\Delta\phi = 1.8°$ in latitude. Richardson tabulated his
initial values for a selection of points over central Europe. The area is shown on a
map on page 184 of *WPNP* (reproduced as Fig. 6.9).

In §9/1 of *WPNP*, Richardson describes the various steps he took in preparing
his initial data. He prepared the mass and wind analyses independently (today, this
is called univariate analysis). The data were obtained from the compilations of
Hergesell and the aerological charts of Bjerknes. The pressure values were computed from heights read directly from Bjerknes' charts. The momentum values were

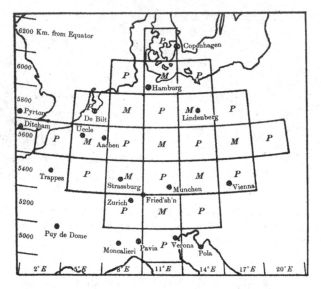

Figure 6.9 Grid used by Richardson for his forecast. The pressure was specified at points denoted P, and the momentum at points denoted M. The prediction was confined to the calculation of the initial tendencies at the P-point near Munich (München) and the M-point directly to the north.

computed using the observations tabulated by Hergesell, followed by visual interpolation or extrapolation to the grid points. Richardson recognised the uncertainty of this procedure: 'It makes one wish that pilot balloon stations could be arranged in rectangular order, alternating with stations for registering balloons . . . ', as depicted on the frontispiece of *WPNP* (see Fig. 1.6 on page 21). In his preface, Richardson acknowledges the substantial assistance of his wife Dorothy in processing the observational data.

The values given by Richardson in his 'Table of Initial Distribution' are reproduced in Table 6.4 on page 115 (the values are converted to modern units). For the white cells, the pressure (in hPa) at the base of each layer is given. For the black cells, the eastward and northward components of momentum for each of the five layers are tabulated (units $10^2 \times \text{kg m}^{-1}\text{s}^{-1}$). The surface elevation in metres is also given (the bottom number in each cell). The latitude is indicated on the left-hand side and the longitude in the top row. We will compare Richardson's values to the reanalysed values after the method of obtaining the latter is discussed.

6.4.2 Reanalysis of the data

The initial fields used in the present study were obtained from the same data sources as those used by Richardson, but we did not follow his method precisely; the

RICHARDSON GRID

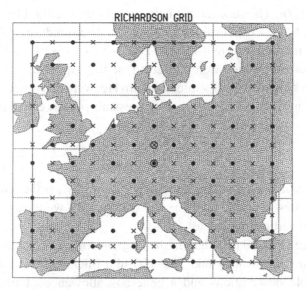

Figure 6.10 The geographical coverage used in repeating and extending Richardson's forecast. P-points are indicated by spots and M-points by crosses. The P-point and M-point for which Richardson calculated his tendencies are shown by encircled marks.

procedure adopted for the re-analysis is outlined below. The geographical coverage used in repeating Richardson's forecast is shown in Fig. 6.10. P-points are indicated by solid circles and M-points by crosses. The region was chosen to best fulfil conflicting requirements: that it be as large as possible; that data coverage over the area be adequate; and that the points used by Richardson be located centrally in the region. The absence of observations precluded the extension of the region beyond that shown. The P-point and M-point for which Richardson calculated his tendencies are shown by encircled marks. In order that the geostrophic relationship should not be allowed to dominate the choice of values, the pressure and velocity analyses were performed separately (and by two different people).

The mass field

The initial pressure fields at level interfaces were derived from Bjerknes' charts of geopotential height at 200, 400, 600 and 800 hPa (his charts 6, 8, 10 and 12). A transparent sheet marked with the grid-points was super-imposed on each chart and the height at each point read off. Each level p_k corresponds to a standard height z_k with temperature T_k. Conversion from height z to pressure p was made using the simple formula

$$p = p_k \left(1 - \frac{z - z_k}{H_k} \right)$$

where $H_k = \Re T_k/g$. The geodynamic heights z_k of the standard levels were 1.959, 4.113, 7.048 and 11.543 km (see *WPNP*, p. 181). The standard temperatures T_k at the surface and interfaces were $+15°C$, $+2°C$, $-12°C$, $-32°C$ and $-50°C$.

Sea-level pressure values were extracted in the same way as heights, from Bjerknes' Chart #1 (Fig. 6.5). His values, in mm Hg, were converted to hectopascals by multiplication by 4/3. Then the surface pressure p_S was calculated from

$$p_S = p_{SEA}\left(1 - \frac{\gamma h}{T_0}\right)^{g/\gamma \Re}$$

where h is orographic height at the point in question, p_{SEA} is the sea-level pressure and standard values $T_0 = 288$ K and $\gamma = 0.0065$ K m^{-1} were used for the surface temperature and vertical lapse-rate.

In the absence of a 100 hPa chart in the Leipzig collection for the time in question, the 100 hPa topography and 200–100 hPa thickness were analysed using the few available observations and a generous allowance of imagination. The thickness values $\Delta z = z_{100} - z_{200}$ were then used to calculate the stratospheric temperature,

$$T_1 = \left(\frac{g\bar{p}}{\Re \Delta p}\right)\Delta z,$$

where $\bar{p} = 150$ hPa and $\Delta p = 100$ hPa are the mean pressure and pressure thickness of the layer. Considering the uncertainties, the values were surprisingly close to those obtained by Richardson (see below).

The momentum field

The initial values of momenta for each of the five layers are required. These were derived from the wind velocities at the intermediate levels 100, 300, 500, 700 and 900 hPa. The observed wind speeds and directions for each level, as compiled by Hergesell and also tabulated in Bjerknes' publication, were plotted on charts upon which isotachs and isogons (lines of constant wind speed and direction) were then drawn by hand. The grid-point values of speed and direction were then read off. It was necessary to exercise a degree of imagination as the observational coverage was so limited, particularly over the Iberian peninsula. The wind values were converted to components u and v and the layer momenta U and V were defined by

$$U = Ru = \frac{\Delta p}{g}v, \qquad V = Rv = \frac{\Delta p}{g}v \tag{6.1}$$

where Δp is the pressure across the layer (obtained in the pressure analysis).

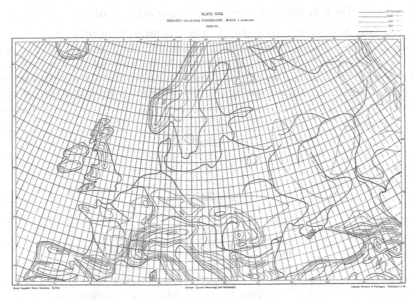

Figure 6.11 Bjerknes' chart of 'greatly idealised' orography. Contours are at 200 m (dotted), 500 m (dashed), 1000 m (solid) and 2000 m (solid). (Plate xxix in Bjerknes *et al.*, 1911)

Orography

The atlas included with Part II of *Dynamic Meteorology and Hydrography* (Bjerknes *et al.*, 1911) contains two charts of orographic height, one 'moderately idealised' and one 'greatly idealised'. Values of surface height at each grid-point were read off from the latter chart (Bjerknes *et al.*, 1911, Plate XXIX), which is reproduced in Fig. 6.11. As Richardson remarks, 'At some points there is a large uncertainty as to the appropriate value of h; for example in Switzerland the uncertainty amounts to several hundred metres.'

6.4.3 Tables of initial data

The pressure, temperature and momentum values, at a selection of points in the centre of the domain, resulting from the re-analysis, are given in Table 6.3. The corresponding values obtained and used by Richardson, extracted from his 'Table of Initial Distribution'(*WPNP*, p. 185), are reproduced in Table 6.4. The orographic heights are also indicated (bottom number in each block). To facilitate comparison, the orography values used by Richardson were also used in the re-analysis.

There is reasonable agreement between the pressure and stratospheric temperature values in the two tables. In general, pressure differences are within one or two hectopascals. There is a notable exception at the point (48.6°N, 5.0°E), where the

Table 6.3 *Initial distribution: reanalysed values.*

The pressure (units hPa) at the base of each of the five layers is given for the white cells. The stratospheric temperature (units K) is also given. The eastward and northward components of momentum (units $10^2 \times \mathrm{kg\,m^{-1}s^{-1}}$) for each of the five layers are tabulated in the black cells. The bottom number in each cell is the surface elevation (in metres). Latitude is indicated on the left-hand side and longitude in the top row.

	5°E	8°E	11°E	14°E	17°E
54.0°N			106 −228 120 −144 0 −81 −97 0 −221 81 0		
52.2°N		−62 −138 −133 −135 −155 150	206 410 609 799 987 200	−25 −79 −107 −156 −181 100	
50.4°N	−175 208 −292 263 −249 174 −118 99 −88 51 200	216 205 409 607 796 983 200	−105 −182 −268 −38 −201 −18 −199 73 −127 73 400	212 206 410 608 798 976 300	−126 −218 −167 −213 −155 −130 −214 0 −175 82 300
48.6°N	221 204 406 605 793 984 200	−159 −275 −216 −131 −60 400	214 206 410 608 796 961 400	−131 −205 −147 −129 −81 400	213 206 410 608 798 989 200
46.8°N		217 204 405 604 794 872 1200	−208 18 −289 0 −172 172 −45 64 −32 38 1800	213 205 407 606 796 842 1500	
45.0°N			210 203 404 603 795 995 100		

Table 6.4 *Initial distribution: Richardson's values.*

The pressure (units hPa) at the base of each of the five layers is given for the white cells. The stratospheric temperature (units K) is also given. The eastward and northward components of momentum (units $10^2 \times \mathrm{kg\,m^{-1}s^{-1}}$) for each of the five layers are tabulated in the black cells. The bottom number in each cell is the surface elevation (in metres). Latitude is indicated on the left-hand side and longitude in the top row.

	5°E	8°E	11°E	14°E	17°E
54.0°N			−65 8 127 −104 81 −25 −81 0 −198 84 0		
52.2°N		−70 −62 −114 −91 −160 150	214 205 409 609 798 988 200	−160 40 −60 −60 −219 100	
50.4°N	−30 −110 −245 300 −223 158 −91 87 −18 15 200	212 205 408 607 795 983 200	−56 −18 −146 −62 −95 29 −52 58 −110 55 400	214 205 409 609 798 976 300	−100 −32 0 −260 −55 −135 −25 48 −190 160 300
48.6°N	214 203 405 604 793 974 200	27 −328 −136 −33 48 400	212 205 409 608 796 963 400	0 −166 −95 −19 −65 400	214 204 408 607 798 988 200
46.8°N		214 204 406 605 795 875 1200	−50 80 −280 41 −175 150 −105 80 −155 40 1800	214 204 408 607 797 846 1500	
45.0°N			213 203 403 603 796 997 100		

old and new values differ by 10 hPa. We will see below that Richardson's value at this point is suspect. A similar table of initial values appears in Platzman (1967) in which two surface pressures at 46.8°N are question-marked. In fact, it is the orographic heights in Platzman's table that are incorrectly transcribed from *WPNP*.

Comparing the momenta in Tables 6.3 and 6.4, we see much more significant discrepancies. Although the overall flow suggested by the momenta is similar in each case, point values are radically different from each other, with variations as large as the values themselves and occasional differences of sign. These dissimilarities arise partly from the different analysis procedures used, but mainly from the large margin of error involved in the interpolation from the very few observations to the grid-points.

The investigation of the comparison between the old and new momentum values, using modern techniques of objective analysis, is an attractive possibility, but it will not be explored here.[4] Instead, in the forecast in Chapter 7, we have simply replaced the re-analysed values of all fields by Richardson's original values at the (few) gridpoints where the latter are available. The values in Table 6.4 are thus the initial values for both Richardson's forecast and the forecasts described in the following chapters. In consequence of this, the calculated tendencies at the central points of the domain will be found to be essentially the same as those obtained by Richardson.

---

[4] Recently, the European Centre for Medium-Range Weather Forecasts (ECMWF) has completed a re-analysis back to 1957 (see §11.4 below). Perhaps some day 'in the dim future', they will reach back to 1910.

7

Richardson's forecast

Starting from the table of the initially observed state of the atmosphere
... the rates of change of the pressures, temperatures, winds, etc. are
obtained. Unfortunately this 'forecast' is spoilt by errors in the initial
data for winds.   (*WPNP*, p. 2)

We now come to the 'forecast' made by Richardson. We examine what he actually
predicted and consider the reasons why his results were so unrealistic. Richardson
was acutely aware of the shortcomings of his results. In his summary (*WPNP*,
Chapter 1) he wrote that errors in the wind data spoilt the forecast. On page 183, he
described the computed surface pressure tendency as 'absurd' and, four pages later,
called it a 'glaring error'. Yet he believed that the cause of the outrageous results
was not the forecasting method itself but the initial data and that, with appropriate
smoothing of the data, it would be reasonable to expect the forecast to agree with
the actual smoothed weather (*WPNP*, p. 217).

   In §7.1 we review the nature and extent of Richardson's prediction. Considerable
insight into the causes of the forecast failure comes from a scale analysis of the
equations, in which the relative sizes of the various terms are examined and compared. This is undertaken in §7.2. The initial tendencies that comprise the forecast
are then analysed in §7.3. In §7.4, we begin to examine Richardson's discussion of
the causes of his forecast failure. Finally, we review the work of Margules, published almost twenty years before *WPNP*, which pointed to serious problems with
Richardson's methodology.

7.1 What Richardson actually predicted: 20 numbers

Richardson's forecast amounted to the calculation of *20 numbers*, the changes of
the components of horizontal momentum at an M-point and of pressure, humidity
and stratospheric temperature at a P-point. The essentials of the computational

117

Table 7.1 *Initial values for the two prediction cells.*

P-point: pressure p (hPa) at the base of each layer, temperature T (K) for the stratosphere and water content W (kg m^{-2}) for the lower layers. M-points: eastward and northward components of momentum U and V (kg m^{-1} s^{-1}) for each layer.

P-point		M-point	
$p_1 = 205.0$	$T_1 = 212$	$U_1 = -5600$	$V_1 = -1800$
$p_2 = 409.0$	$W_2 = 0.0$	$U_2 = -14\,600$	$V_2 = -6200$
$p_3 = 607.9$	$W_3 = 1.0$	$U_3 = -9500$	$V_3 = +2900$
$p_4 = 796.0$	$W_4 = 4.0$	$U_4 = -5200$	$V_4 = +5800$
$p_S = 962.6$	$W_5 = 9.0$	$U_5 = -11\,000$	$V_5 = +5500$

algorithm were given in §5.6 above. A complete appreciation of Richardson's actual computations would require a detailed analysis of his 23 computing forms. The crucial role of these forms was emphasised by Richardson: 'The computing forms ... may be regarded as embodying the process and therefore summarizing the whole book' (*WPNP*, p. 181). We leave such a detailed analysis to future investigators and confine our attention here to the primary results obtained by Richardson.

The initial data for the forecast were presented in Table 6.4 on page 115 above. The prognostic variables were the two components of momentum, U and V, and the pressure p, specified on alternate checkers of a chessboard grid, with five values of each variable in each vertical column. The temperature of the uppermost stratum was also given at P-points, as was the total mass of water in each of the lower four strata. There were thus ten prognostic variables at P-points and ten at M-points, whose initial values are given in Table 7.1 (all values are in SI units). To compute the rates of change of the 20 quantities, their values in a small neighbourhood surrounding the central points were required. It is these values that are presented in Richardson's 'Table of Initial Distribution', on page 185 of *WPNP*, and reproduced (except for the moisture variable) in Table 6.4 above. Richardson applied his numerical process to these initial data, using the computing forms to organise the calculations. All calculations were performed twice and the results compared to avoid or at least reduce the likelihood of errors. Multiplications were done using a 25 cm slide rule. This gave an accuracy to three significant digits. For more sensitive terms, such as the pressure gradients (see Eq. (5.22) on page 94), five-figure logarithms were found to be essential.

The set of 23 computing forms was in two groups, with 19 for P-points and four for M-points (see Table 1.3 on p. 24). The P-point selected for the forecast was at 48.6°N, 11°E. This point is in Bavaria, about 65 km north-west of Munich. The M-point was 200 km to the north (at 50.4°N, 11°E). The vertical velocity,

Table 7.2 *Richardson's 'forecast': predicted six-hour changes in pressure* Δp,
stratospheric temperature ΔT *and water content* ΔW *at the P-point and
eastward and northward components of momentum* ΔU *and* ΔV *at the M-point.*
The variables and units are the same as in Table 7.1 above.

P-point		M-point	
$\Delta p_1 = 48.3$	$\Delta T_1 = 19.6$	$\Delta U_1 = -73\,000$	$\Delta V_1 = -33\,700$
$\Delta p_2 = 77.0$	$\Delta W_2 = 0.07$	$\Delta U_2 = -19\,600$	$\Delta V_2 = +23\,800$
$\Delta p_3 = 103.2$	$\Delta W_3 = 0.24$	$\Delta U_3 = -8900$	$\Delta V_3 = +13\,800$
$\Delta p_4 = 126.5$	$\Delta W_4 = 1.49$	$\Delta U_4 = -15\,300$	$\Delta V_4 = -4300$
$\Delta p_S = 145.1$	$\Delta W_5 = 4.02$	$\Delta U_5 = -17\,900$	$\Delta V_5 = +6300$

which is computed at P-points, is required also at M-points. Ideally, it would have
been computed there by interpolation. However, in view of the limited nature of
the forecast, the values of w at the P-point were also used by Richardson in the
momentum equations at the M-point.

The calculated tendencies are tabulated on page 211 of *WPNP*, expressed as
changes over a six-hour period centred on the initial time. For example, the computed value of surface pressure tendency $\partial p_S / \partial t$ was multiplied by 21 600 seconds
to estimate the change in surface pressure between 0400 UTC and 1000 UTC. The
changes are given in Table 7.2. An example of Richardson's wry sense of humour
is given by the running head on the page where the forecast values are tabulated;
it reads: THE RESULTING 'PREDICTION'. It is important to stress that Richardson's
forecast was confined to the 20 numbers in this table.

The result most frequently quoted from *WPNP* is the predicted change in surface
pressure of 145 hPa in six hours. However, the other forecast changes are also problematical. Richardson wrote below his table of results: 'It is claimed that the above
form a fairly correct deduction from a somewhat unnatural initial distribution.' This
remarkable claim will be considered in depth below.

A computer program has been written to repeat and extend Richardson's forecast,
using the same initial values. We will describe this in more detail in Chapter 9 below.
For now, we just note that the computer model produces results consistent with those
obtained manually by Richardson. In particular, the 'glaring error' in the surface
pressure tendency is reproduced almost exactly by the model. These results confirm
the essential correctness of Richardson's calculations. Such manual calculations,
done with slide-rule and logarithm tables, are notoriously error-prone. Although a
number of minor slips were detected in Richardson's calculations, they did not have
any significant effect on his results. We may conclude that his unrealistic results
were not due to arithmetical blunders, but had a deeper cause.

Richardson's calculations involved an enormous amount of numerical computation. Despite the limited scope of his forecast, it cost him some two years of arduous calculation (see Appendix 4). Moreover, the work was carried out in the Champagne district of France where Richardson served as an ambulance driver during World War I:

> ...the detailed example of Ch. IX was worked out in France in the intervals of transporting wounded in 1916–1918. During the Battle of Champagne in April, 1917 the working copy was sent to the rear, where it became lost, to be re-discovered some months later under a heap of coal.   *(WPNP, preface)*

Richardson described his working conditions with characteristic stoicism: 'My office was a heap of hay in a cold rest billet' (*WPNP*, p. 219). His dedication and tenacity in the dreadful circumstances of trench warfare should serve as an inspiration to those of us who work in more congenial conditions.

7.2 Scaling the equations of motion

7.2.1 Richardson's scaling

Although Richardson was reluctant to make any unnecessary approximations in the equations, he was forced by circumstances to replace the equation for vertical acceleration by the hydrostatic balance relation. He could have introduced a number of other simplifications without serious error, but argued that, since the numerical process could handle terms with ease irrespective of their size, they might as well be retained. He did investigate the relative sizes of the terms in the continuity and momentum equations, and set down estimates of the extreme values ordinarily attained by the various terms:

> These figures have been obtained by a casual inspection of observational data and they may be uncertain except as to the power of ten. They are expressed in C.G.S. units. They relate only to the large-scale phenomena that can be represented by the chosen co-ordinate differences of 200 kilometres horizontally, one fifth of the pressure vertically and by the time-step of six hours.   *(WPNP, p. 22)*

The sizes assigned by Richardson to the terms of the continuity equation are given here (they are in SI units, and we omit a small term, as discussed in Chapter 2):

$$\underbrace{\frac{\partial \rho}{\partial t}}_{10^{-6}} + \underbrace{\frac{\partial \rho u}{\partial x}}_{10^{-4}} + \underbrace{\frac{\partial \rho v}{\partial y}}_{10^{-4}} - \underbrace{\frac{\rho v \tan \phi}{a}}_{10^{-5} \tan \phi} + \underbrace{\frac{\partial \rho w}{\partial z}}_{<10^{-4}} = 0 \qquad (7.1)$$

Clearly, he was unsure of the appropriate magnitude for vertical momentum. We note a crucial implication of the scaling: the tendency term in (7.1) is two orders of magnitude smaller than the largest terms in the equation, so that the density at a fixed point is close to constant. As Bjerknes and others had found earlier, this is a better approximation than assuming the atmosphere behaves like a liquid; that is,

$$\nabla \cdot \rho \mathbf{v} = 0 \qquad \text{is better than} \qquad \nabla \cdot \mathbf{v} = 0.$$

The quasi-nondivergence of specific momentum was the primary reason it was chosen by Richardson as a dependent variable, in preference to velocity (*WPNP*, p. 24).

The terms of the horizontal equations of motion in flux form were assigned magnitudes as indicated here:

$$\underbrace{\frac{\partial \rho u}{\partial t}}_{10^{-3}} + \underbrace{\frac{\partial \rho u^2}{\partial x}}_{10^{-3}} + \underbrace{\frac{\partial \rho u v}{\partial y}}_{10^{-2}} + \underbrace{\frac{\partial \rho u w}{\partial z}}_{10^{-2}?} - \underbrace{f\rho v}_{10^{-2}} + \underbrace{\frac{2\rho u v \tan \phi}{a}}_{10^{-3}\tan \phi} + \underbrace{\frac{\partial p}{\partial x}}_{10^{-2}} = 0 \quad (7.2)$$

$$\underbrace{\frac{\partial \rho v}{\partial t}}_{10^{-3}} + \underbrace{\frac{\partial \rho v u}{\partial x}}_{10^{-3}} + \underbrace{\frac{\partial \rho v^2}{\partial y}}_{10^{-2}} + \underbrace{\frac{\partial \rho v w}{\partial z}}_{10^{-2}?} + \underbrace{f\rho u}_{10^{-2}} + \underbrace{\frac{(\rho u^2 - \rho v^2) \tan \phi}{a}}_{10^{-3}\tan \phi} + \underbrace{\frac{\partial p}{\partial y}}_{10^{-2}} = 0 \quad (7.3)$$

(an error in Richardson's scaling of the geometric term in (7.3) has been amended). Again, we note the uncertainty regarding terms with vertical momentum. The non-linear terms involving zonal gradients were assigned smaller scales than those involving meridional gradients. This assumption was presumably inspired by Richardson's 'casual inspection of observational data'. However, despite the pre-dominantly zonal character of the mid-latitude flow, the eastward flux is generally comparable in magnitude to the northward flux.

The sizes of the terms assigned by Richardson are not typical, but represent the extreme values ordinarily attained by the terms. They are such that the nonlinear terms are comparable to the Coriolis and pressure gradient terms, so that geostrophic balance does not obtain. We may set the horizontal and vertical scales $\mathcal{L}$ and $\mathcal{H}$ and the time scale $\mathcal{T}$ to represent synoptic variations:

$$\mathcal{L} = 10^6 \, \text{m}, \qquad \mathcal{H} = 10^4 \, \text{m}, \qquad \mathcal{T} = 10^5 \, \text{s}.$$

Then Richardson's values are consistent with the following scales for the dependent variables:

$$\mathcal{V} = 10^2 \, \text{m s}^{-1}, \qquad \mathcal{W} \leq 1 \, \text{m s}^{-1}, \qquad \mathcal{P}' = 10^4 \, \text{Pa}, \qquad \mathcal{R}' = 10^{-1} \, \text{kg m}^{-3}.$$

Here, $\mathcal{V}$ and $\mathcal{W}$ are the scales of the horizontal and vertical motion and $\mathcal{P}'$ and $\mathcal{R}'$ are the scales of p' and ρ', the deviations from the mean values $\bar{p}(z)$ and $\bar{\rho}(z)$. Indeed, the chosen scales represent very active atmospheric conditions, more typical of extremes than of average synoptic patterns. The pressure and density variations amount to 10 per cent of their mean values. Moreover, the scale of vertical velocity is unrealistically large (Richardson's question-mark confirms his doubt). Assuming $w \ll 1 \, \mathrm{m\,s^{-1}}$, the dominant terms of the momentum equations yield the well-known diagnostic relationship for the *gradient wind*:

$$\mathbf{v}.\nabla\mathbf{v} + f\mathbf{k} \times \mathbf{v} + (1/\rho)\nabla p = 0.$$

This is an extension of geostrophic balance in which the curvature of the flow is taken into account through the advection term.

It is clear from the foregoing that Richardson had an excellent understanding of the concept of scale analysis and a refined capacity to employ it to effect. It is unfortunate that he did not pursue this avenue, as it might have led him to methods of greater utility than the sledge-hammer approach of keeping all terms. Platzman (1967, p. 530) expressed a similar view when he wrote that Richardson's disregard of perturbation theory as a means of clarifying the problems of dynamic meteorology was a major defect in his approach to weather prediction. Scale analysis was the basis of the development of the quasi-geostrophic system some decades later (Charney, 1948).

7.2.2 Re-scaling using typical synoptic values

To facilitate further analysis, it is convenient to re-scale the equations using typical synoptic values rather than Richardson's extreme values. We assume characteristic sizes for the independent variables as before, and set the following scales for the dependent variables:

$$\mathcal{V} = 10 \, \mathrm{m\,s^{-1}}, \quad \mathcal{W} = 10^{-1} \, \mathrm{m\,s^{-1}}, \quad \mathcal{P}' = 10^3 \, \mathrm{Pa}, \quad \mathcal{R}' = 10^{-2} \, \mathrm{kg\,m^{-3}}.$$

Now the continuity equation scales as follows:

$$\underbrace{\frac{\partial \rho}{\partial t}}_{10^{-7}} + \underbrace{\frac{\partial \rho u}{\partial x}}_{10^{-5}} + \underbrace{\frac{\partial \rho v}{\partial y}}_{10^{-5}} - \underbrace{\frac{\rho v \tan\phi}{a}}_{10^{-6}\tan\phi} + \underbrace{\frac{\partial \rho w}{\partial z}}_{10^{-5}} = 0. \tag{7.4}$$

Considering the gas law, the variations in the mass variables are related by

$$p' = \Re(\bar{T}\rho' + \bar{\rho}T'). \tag{7.5}$$

Then, noting that $\Re \approx 300 \, \mathrm{m^2\,s^{-2}\,K^{-1}}$, and $\bar{T} \approx 300 \, \mathrm{K}$, the pressure tendency may be estimated:

$$\frac{\partial p}{\partial t} \approx \Re\bar{T}\frac{\partial \rho}{\partial t} \approx 10^{-2} \, \mathrm{Pa\,s^{-1}}. \tag{7.6}$$

A practical unit for pressure tendency, which is more 'ergonomic' than $\mathrm{Pa\,s^{-1}}$, has been introduced by Sanders and Gyakum (1980) in a study of rapidly intensifying

oceanic cyclones (which they called 'bombs'). A deepening rate of 24 hPa in
24 hours (at 60°N) was defined to be one Bergeron (1 Ber) [named for the renowned
Swedish meteorologist Tor Bergeron]. Here we define 1 Ber to be a tendency of *one
hectopascal per hour*. Thus, using (7.6), the typical synoptic scale of the pressure
tendency is about 10^{-4} hPa s^{-1}, or 0.36 Ber. Richardson's prediction of 145 hPa
in six hours is 24 Ber, about two orders of magnitude larger than typical synoptic
values.

The horizontal equations of motion, using typical synoptic scales, are as follows:

$$\underbrace{\frac{\partial \rho u}{\partial t}}_{10^{-4}} + \underbrace{\frac{\partial \rho u^2}{\partial x}}_{10^{-4}} + \underbrace{\frac{\partial \rho u v}{\partial y}}_{10^{-4}} + \underbrace{\frac{\partial \rho u w}{\partial z}}_{10^{-4}} - \underbrace{f\rho v}_{10^{-3}} + \underbrace{\frac{2\rho u v \tan \phi}{a}}_{10^{-5}\tan\phi} + \underbrace{\frac{\partial p}{\partial x}}_{10^{-3}} = 0 \quad (7.7)$$

$$\underbrace{\frac{\partial \rho v}{\partial t}}_{10^{-4}} + \underbrace{\frac{\partial \rho v u}{\partial x}}_{10^{-4}} + \underbrace{\frac{\partial \rho v^2}{\partial y}}_{10^{-4}} + \underbrace{\frac{\partial \rho v w}{\partial z}}_{10^{-4}} + \underbrace{f\rho u}_{10^{-3}} + \underbrace{\frac{(\rho u^2 - \rho v^2)\tan \phi}{a}}_{10^{-5}\tan\phi} + \underbrace{\frac{\partial p}{\partial y}}_{10^{-3}} = 0 \quad (7.8)$$

(we do not distinguish between the scales of zonal and meridional fluxes). The
Coriolis and pressure gradient terms now dominate everything else, and this fundamental balance corresponds to *quasi-geostrophic flow*:

$$f\mathbf{k} \times \mathbf{v} + \frac{1}{\rho}\nabla p = 0 \qquad \text{or} \qquad \mathbf{v} = \frac{1}{f\rho}\mathbf{k} \times \nabla p.$$

It must not be inferred from the re-scaling that Richardson's choice of scales was
incorrect or inappropriate. The important point is that he chose values characteristic
of atmospheric extremes rather than of typical synoptic conditions. This choice
may have made it more difficult for him to develop approximations applicable to
more moderate conditions. We saw in Chapter 6 that the atmospheric conditions
prevailing on the occasion for which he made his forecast were quite temperate,
with little change over the observation period.

7.2.3 The triple compensation effect

How can the tendencies be small when they are determined by quantities that
are orders of magnitude larger? The answer is that, for normal atmospheric conditions, there are several compensating effects that result in strong cancellation
between terms. The most prominent is the balance between the Coriolis and pressure gradient terms, corresponding to quasi-geostrophic flow. Geostrophic flow is
also quasi-nondivergent, its divergence depending only on the gradient of planetary
vorticity, the β-effect. In general, the atmospheric divergence is small relative to its
components. The pressure tendency equation follows from integration through the

atmosphere of the continuity equation (7.4):

$$\frac{\partial p_S}{\partial t} = -\int_0^{p_S} \frac{1}{\rho} \nabla \cdot \rho \mathbf{v} \, dp. \tag{7.9}$$

With naive synoptic scaling, the individual components of the integrand are $O(10^{-5})$, so the right-hand side appears to be $O(1)$, or about 36 Ber, comparable to Richardson's computed tendency. But this is not the case, because of the triple compensation effect:

(i) *Horizontal compensation*: horizontal influx and out-flow have a strong tendency to offset each other so that the horizontal divergence is smaller than its component terms.

(ii) *Vertical compensation*: horizontal influx at a point tends to correspond to vertical out-flow, and vice versa. Thus, the three-dimensional divergence is characteristically smaller than the two-dimensional divergence.

(iii) *Dines compensation*: inflow at low levels and out-flow at higher levels of a column, or vice versa, tend to occur simultaneously, resulting in cancellation when the vertical integral is taken.

The first two effects moderate the change in pressure difference across a layer. The first and third act to reduce the change in pressure through a vertical column, that is, the surface pressure tendency. The net effect is that typical values of pressure tendency are about one hundredth of the size that a naive scaling argument might suggest.

7.2.4 The effect of data errors

The quasi-geostrophic and quasi-nondivergent nature of the atmospheric flow are delicate, depending for their persistence on effects that closely cancel each other. If the analysed state were error-free, it would reflect this balance. However, this ideal is never attained, and observational errors can have drastic consequences. Suppose there is a ten per cent error in the v-component of the wind observation at a specific point, $\Delta v \approx 1 \, \mathrm{m \, s^{-1}}$. The scales of the terms in the u-equation are as before:

$$\underbrace{\frac{\partial \bar{\rho} u}{\partial t}}_{10^{-4}} - \underbrace{f \bar{\rho} (v + \Delta v)}_{10^{-3}} + \underbrace{\frac{\partial p}{\partial x}}_{10^{-3}} = 0 \tag{7.10}$$

(for simplicity, we omit nonlinear terms). However, the error in the tendency is $\Delta(\partial \bar{\rho} u / \partial t) \sim f \bar{\rho} \Delta v \sim 10^{-4} \, \mathrm{kg \, m^{-2} \, s^{-2}}$, comparable in size to the tendency itself: the signal-to-noise ratio is 1 (the symbol '$\sim$' represents equality of order of magnitude). The forecast may be qualitatively reasonable, but it will be quantitatively invalid.

A graver conclusion is reached for a ten per cent error $\Delta p \sim 1$ hPa in the pressure variation: the numerical gradient is not computed using the horizontal scale length $\mathcal{L}$, but the grid scale Δx which is much smaller (say, $\Delta x = \mathcal{L}/10$). Consequently, the error in its gradient is correspondingly large:

$$\frac{\partial p}{\partial x} \sim \frac{\mathcal{P}'}{\mathcal{L}}, \qquad \text{but} \qquad \Delta \frac{\partial p}{\partial x} \sim \frac{\Delta p}{\Delta x} \sim \frac{\mathcal{P}'}{\mathcal{L}} \sim \frac{\partial p}{\partial x},$$

so the error is comparable to the term itself and the error in the wind tendency is now

$$\Delta \frac{\partial \bar{\rho} u}{\partial t} \sim \frac{\partial p}{\partial x} \sim 10^{-3} \gg \frac{\partial \bar{\rho} u}{\partial t}.$$

That is, the error in momentum tendency is larger than the tendency itself; the forecast will be qualitatively incorrect, indeed unreasonable.

Now consider the continuity equation. As we have seen, the pressure tendency has characteristic scale 10^{-2} Pa s^{-1} or 0.36 Ber, corresponding to a scale for the tendency of density of

$$\frac{\partial \rho}{\partial t} \sim 10^{-7} \, \text{kg m}^{-3} \, \text{s}^{-1}.$$

The mass divergence $D = \nabla \cdot \rho \mathbf{v}$ is $O(10^{-7})$ but its individual components are $O(10^{-5})$. If there is a ten per cent error in the northward wind component v, naive scaling would suggest that the resulting error in divergence is $\Delta D \sim \Delta v/\mathcal{L} \sim 10^{-6}$. This error is larger than the divergence itself! As a result, the tendency of density is unrealistic. But matters are worse still: since the numerical divergence is computed using the grid scale Δx, the error is correspondingly greater:

$$\Delta D \sim \Delta \frac{\partial v}{\partial x} \sim \frac{\Delta v}{\Delta x} \sim \frac{\mathcal{V}}{\mathcal{L}} \sim 10^{-5} \sim 10^2 \times D.$$

This yields a tendency *two orders of magnitude too large*. By (7.6), this also implies a pressure tendency two orders of magnitude larger than the correct value. Instead of the value $\partial p/\partial t \sim 0.36$ Ber, we get a change of order 36 Ber. This is strikingly reminiscent of Richardson's result.

7.3 Analysis of the initial tendencies

Richardson ascribed the unrealistic value of pressure tendency to errors in the observed winds, which resulted in spuriously large values of calculated divergence. This is true as far as it goes. However, the problem is deeper: even if the winds were modified to remove divergence completely at the initial time, large tendencies would soon be observed. The close balance between the pressure and wind fields, and the compensations that result in quasi-nondivergence, ensure that the high-frequency gravity waves have much smaller amplitude than the rotational part of the flow, and that the tendencies are small. As we have seen from scale analysis, minor errors in observational data can result in a disruption of the balance, and cause spuriously large tendencies. The imbalance engenders large-amplitude gravity wave oscillations. They can be avoided by modifying the data to restore harmony between the fields. In Chapter 8 we will describe several methods of achieving balance and in Chapter 9 we will apply a simple initialisation method to Richardson's data and show that it yields realistic results. Here we examine the initial data used for the forecast and show that they are far from balance.

7.3.1 The divergence and the pressure tendency

Platzman (1967) examined Richardson's results and discussed two problems contributing to the large pressure tendency: the horizontal divergence values are too large, due to lack of cancellation between the terms, and there is a lack of compensation between convergence and divergence in the vertical. We follow the same approach, developing and expanding Platzman's analysis. The Computing Form P_{XIII}, reproduced in Table. 7.3, contains the results of Richardson's calculations for pressure changes. Table 7.4, constructed using values from Computing Form P_{XIII}, shows the total divergence and its components. The eastward and northward components are given in columns 2 and 3 and their sum, the horizontal divergence in column 4. The numbers in the bottom row of the table are the averages of the layer absolute values, and indicate the characteristic magnitudes of the terms. We see that the horizontal divergence is comparable in size to its components: there is no horizontal compensation. The vertical component of divergence (column 5) is comparable in size to the horizontal, but there is only a slight tendency for these terms to cancel: the vertical compensation is weak. Finally, column 6 of the table shows that the divergence is negative in all layers: the Dines compensation mechanism is completely absent from the data.

Table 7.5 shows the change in pressure thickness for each layer (column 4) and the change in pressure at the base of each layer (column 5). As a result of convergence in all layers, the change in the pressure difference across all layers is positive, and the cumulative effect is a huge pressure rise at the surface.

7.3.2 The momentum tendencies

For balanced atmospheric flow, the tendencies arise as small residual differences between large quantities. But this is not the case for Richardson's data. The calculated changes in momentum appear on the bottom rows of Computing Forms

Table 7.3 *Richardson's Computing Form* P$_{XIII}$, *giving the divergence of momentum (in column 5) and the pressure changes (last column).*

The figure in the bottom right corner is the much-discussed forecast change in surface pressure.

COMPUTING FORM P XIII. Divergence of horizontal momentum-per-area. Increase of pressure

The equation is typified by : $-\dfrac{\partial R_{EE}}{\partial t} = \dfrac{\partial M_{EE}}{\partial e} + \dfrac{\partial M_{NE}}{\partial n} - M_{NE}\dfrac{\tan\phi}{a} + m_E - m_{EE}* + \dfrac{2}{a}M_{EE}.$ (See Ch. 4/2#5.)

* In the equation for the lowest stratum the corresponding term $- m_{ea}$ does not appear

	Longitude 11° East $\delta e = 441 \times 10^5$			Latitude 5400 km North $\delta n = 400 \times 10^5$			Instant 1910 May 20^d 7^h G.M.T. $a^{-1}.\tan\phi = 1.73\times10^{-9}$		Interval, δt 6 hours $a = 6.36\times10^8$		
REF.:—				previous 3 columns	previous column		Form P XVI	Form P XVI	equation above	previous column	previous column	previous column
h	$\dfrac{\partial M_E}{\partial e}$	$\dfrac{\partial M_N}{\partial n}$	$-\dfrac{M_N\tan\phi}{a}$	div$'_{EN}M$	$-g\delta t$ div$'_{EN}M$		m_E	$\dfrac{2M_E}{a}$	$-\dfrac{\partial R}{\partial t}$	$+\dfrac{\partial R}{\partial t}\delta t$	$g\dfrac{\partial R}{\partial t}\delta t$	$\dfrac{\partial p}{\partial t}\delta t$
	$10^{-5}\times$	$10^{-5}\times$	$10^{-5}\times$	$10^{-5}\times$	$100\times$		$10^{-5}\times$	$10^{-5}\times$	$10^{-5}\times$		$100\times$	$100\times$
h_0							0					0
	-61	-245	-6	-312	656				-229	49·5	483	
h_2							-83					483
	367	-257	2	112	-236			0·06	-136	29·4	287	
h_4							165					770
	93	-303	-16	-226	478			0·11	-124	26·8	262	
h_6							63					1032
	32	-55	-12	-35	74			0·07	-110	23·8	233	
h_8							138					1265
	-256	38	-8	-226	479			0·03	-88	19·0	186	
h_9												1451

Leave the subsequent columns to be filled up after the vertical velocity has been computed on Form P XVI

NOTE: div$'_{EN}M$ is a contraction for
$\dfrac{\delta M_E}{\delta e} + \dfrac{\delta M_N}{\delta n} - M_N\dfrac{\tan\phi}{a}$

SUM =
1451
$= \dfrac{\partial p_a}{\partial t}\delta t$

check by
$\Sigma\; g\delta t$div$'_{EN}M$

Table 7.4 *Analysis of the components of the mass divergence in each layer and the mean absolute values for the column.*

$[\rho w]$ denotes a vertical difference across the layer. All values are in SI units (kg m^{-2} s^{-1}).

Layer	Eastward component $\dfrac{\partial U}{\partial x}$	Northward component $\dfrac{\partial(V\cos\phi)}{\cos\phi\,\partial y}$	Horizontal divergence $\nabla\cdot\mathbf{U}$	Vertical divergence $[\rho w]$	Total divergence $\nabla\cdot\mathbf{U}+[\rho w]$
I	-0.0061	-0.0251	-0.0312	$+0.0083$	-0.0229
II	$+0.0367$	-0.0255	$+0.0112$	-0.0248	-0.0136
III	$+0.0093$	-0.0319	-0.0226	$+0.0102$	-0.0124
IV	$+0.0032$	-0.0067	-0.0035	-0.0075	-0.0110
V	-0.0256	$+0.0030$	-0.0226	$+0.0138$	-0.0088
Mean abs. value	0.0162	0.0184	0.0182	0.0129	0.0137

Table 7.5 *Analysis of the pressure changes (hPa) across each layer, and the pressure change at the base of each layer.*

([p] denotes the vertical pressure difference across a layer). Values in column 4 are the products of those in column 3 by $-g\,\Delta t$ (the time interval is $\Delta t = 21\,600$ s and for simplicity we assume constant acceleration of gravity, $g = 9.778$).

Layer	Level	Total divergence $\nabla \cdot \mathbf{U} + [\rho w]$	Change in pressure thickness $\dfrac{\partial [p]}{\partial t}\Delta t$	Change in base pressure $\dfrac{\partial p}{\partial t}\Delta t$
I		−0.0229	+48.3	
	1			+48.3
II		−0.0136	+28.7	
	2			+77.1
III		−0.0124	+26.2	
	3			+103.2
IV		−0.0110	+23.3	
	4			+126.5
V		−0.0088	+18.6	
	S			+145.1

M_{III} and M_{IV}, reproduced in Table. 7.6. We tabulate the principal terms of the eastward momentum equation in Table 7.7. The pressure gradient and Coriolis terms dominate, but there is little tendency for them to cancel; indeed, for the top two layers they are of the same sign and act to reinforce each other. The numbers in the bottom row of the table are the absolute averages of the layer values. They show that the pressure gradient term is substantially larger than the Coriolis term, and the tendency of U is comparable in size to the pressure term. We conclude that the data are far from geostrophic balance and, as a result, the tendency of momentum is uncharacteristically large. The imbalance is particularly pronounced in the stratosphere, and the momentum change is correspondingly large there. An analysis of the northward momentum equation is shown in Table 7.8. It indicates a similar departure from geostrophic balance and large values for the tendencies.

7.3.3 The moisture and stratospheric temperature tendencies

The predicted changes in the humidity variable are also questionable. For each of the four lower strata, the total mass of water substance per unit area is given in Table 7.1 (p. 118). The saturation values W_{sat} are given on Richardson's Computing Form P_I. For the lowest stratum, $W_5 = 9\,\mathrm{kg\,m^{-2}}$ and $W_{sat} = 20\,\mathrm{kg\,m^{-2}}$. The six-hour change for this stratum, given in Table 7.2, is $\Delta W_5 = 4.02$. At this rate,

Table 7.6 *Richardson's Computing Forms* M_{III}, *for the eastward component of momentum (top), and* M_{IV}, *for the northward component (bottom).*

The values in these tables are in CGS units.

COMPUTING FORM M III. For the Dynamical Equation for the Eastward Component

$$-\frac{\partial M_{\text{Eee}}}{\partial t} = \frac{\partial P_{\text{ee}}}{\partial e} + \left[p\frac{\partial h}{\partial e}\right]_e^* + \frac{\partial}{\partial e}\left(\frac{M_{\text{E}}^2}{R}\right)_{\text{ee}} + \frac{\partial}{\partial n}\left(\frac{M_{\text{E}}M_{\text{N}}}{R}\right)_{\text{ee}} + \left[m_{\text{E}}v_{\text{E}}\right]_e \quad \dagger \quad -2\omega\sin\phi\,M_{\text{Nee}} + 2\omega\cos\phi\,M_{\text{Bee}} + \left(\frac{3M_{\text{E}}M_{\text{B}} - 2M_{\text{E}}M_{\text{N}}\tan\phi}{aR}\right)_{\text{ee}}.$$

* No corresponding term in upper layers.   † A term $[m_{\text{E}}v_{\text{E}}]_e$ absent because ground impervious to wind.

Longitude 11° E	Latitude 5600 km N	2ω sin φ = 1·124 × 10⁻⁴, 2ω cos φ = 0·930 × 10⁻⁴					Instant 1910 May 20ᵈ 7ʰ G.M.T.	δt = 6 hours		
$\frac{\tan\phi}{a}=1\cdot897\times10^{-9}$	$\delta e=425\times10^5$ for δλ=6°	h_0		h_2		h_4		h_6		h_8	h_9
REF.	Term	+	−	+	−	+	−	+	−	+	−
CH. 4/4 #14 {	$\delta P/\delta e$	31·3		4·9		9·4		12·9			215·5
	$[p.\delta h/\delta e]_e$	———		———		———		———		230·8	
	$\frac{\delta}{\delta e}(M_{\text{E}}^2/R)$	0·5			3·4		2·7		0·5	2·6	
	$\frac{\delta}{\delta n}(M_{\text{E}}M_{\text{N}}/R)$	0·2			0·1	1·5		0·6			0·7
Form P XVI	$m_{\text{E}}v_{\text{E}}$ at upper limit			(0·0)		(0·9)		(0·2)			(0·7)
Form P XVI	$-m_{\text{E}}v_{\text{E}}$ at lower limit	(0·0)		(0·9)		(0·2)		(0·7)	.	* no term here	
Form P XVI	$-2\omega\sin\phi.M_{\text{N}}$	2·0		7·0			3·3	6·5			6·2
Form P XVI	$+2\omega\cos\phi.M_{\text{B}}$	(0·01)		(0·02)	.	(0·03)		(0·02)		(0·01)	
Form P XVI	$+3M_{\text{E}}M_{\text{B}}/(aR)$		(0·0)		(0·0)	(0·0008)		(0·0)			(0·0)
	$-2M_{\text{E}}M_{\text{N}}\tan\phi/(aR)$		0·0		0·2		0·1		0·1		0·1
Form M I	viscosity terms		0·0		0·0		0·0	0·2			1·4
Form M II	stratosphere, special	0·2									
	sums + and −	34·0	0·2	12·8	3·7	11·1	7·0	14·4	7·3	232·9	224·6
	$-\partial M_{\text{E}}/\partial t$	+33·8		+9·1		+4·1		+7·1		+8·3	
	$+\delta t.\partial M_{\text{E}}/\partial t$	−730 × 10⁵		−196 × 10⁵		−89 × 10⁵		−153 × 10⁵		−179 × 10⁵	

COMPUTING FORM M IV. For the Dynamical Equation for the Northward Component

$$-\frac{\partial M_{\text{N}}}{\partial t} = -g_{\text{N}}R + \frac{\partial P}{\partial n} + \left[p\frac{\partial h}{\partial n}\right]_e + \frac{\partial}{\partial e}\left(\frac{M_{\text{N}}M_{\text{E}}}{R}\right) + \frac{\partial}{\partial n}\left(\frac{M_{\text{N}}^2}{R}\right) + \left[m_{\text{N}}v_{\text{N}}\right]_e + 2\omega\sin\phi.M_{\text{E}} + \frac{3M_{\text{N}}M_{\text{B}}}{aR} + \frac{\tan\phi\{M_{\text{E}}^2 - M_{\text{N}}^2\}}{aR}.$$

The above equation relates to the stratum h_0 to h_9. For upper strata, omit the term $p\frac{\partial h}{\partial n}$, and subtract a term $m_{\text{N}}v_{\text{N}}$ at the lower boundary.

Longitude 11° E	Latitude 5600 km N	2ω sin φ = 1·124 × 10⁻⁴, 2ω cos φ = 0·930 × 10⁻⁴					Instant 1910 May 20ᵈ 7ʰ G.M.T.	δt = 6 hours		
$\frac{\tan\phi}{a}=1\cdot897\times10^{-9}$	$\delta e=425\times10^5$ for δλ=6°	h_0		h_2		h_4		h_6		h_8	h_9
REF.	Term	+	−	+	−	+	−	+	−	+	−
Table in Ch. 4/4	$-g_{\text{N}}R$	2·7		1·5		0·9		0·4		0·1	
Ch. 4/4 #14 {	$\delta P/\delta n$	24·1			1·7	2·3		7·7		498·3	
	$[p.\delta h/\delta n]_e$	———		———		———		———			487·8
	$\frac{\delta}{\delta e}(M_{\text{N}}M_{\text{E}}/R)$		0·0	4·1		2·5		0·4	.		2·1
	$\frac{\delta}{\delta n}(M_{\text{N}}^2/R)$	0·4		0·5			1·3		0·4	0·8	
Form P XVI	$m_{\text{N}}v_{\text{N}}$ at upper limit			(0·7)		(0·1)	(0·1)			(0·4)	
Form P XVI	$-m_{\text{N}}v_{\text{N}}$ at lower limit	(0·7)		(0·1)			(0·1)		(0·4)	no term here	
	$+2\omega\sin\phi.M_{\text{E}}$		6·3		16·4		10·7	5·8	ʄ		12·4
Form P XVI	$3M_{\text{N}}M_{\text{B}}/(aR)$		(0·0)		(0·0)	(0·0002)		(0·0)		(0·0)	
	$\tan\phi\{M_{\text{E}}^2-M_{\text{N}}^2\}/(aR)$	0·0		0·2		0·1				0·0	0·1
Form M I	viscosity terms		0·0		0·0		0·0		0·0	0·2	
Form M II	stratosphere, special	3·8		———				———			
	sums + and −	26·8	11·2	7·1	18·1	5·8	12·2	8·6	6·6	499·4	502·3
	$-\partial M_{\text{N}}/\partial t$	+15·6		−11·0		−6·4		+2·0		−2·9	
	$+\delta t.\partial M_{\text{N}}/\partial t$	−337 × 10⁵		+238 × 10⁵		+138 × 10⁵		−43 × 10⁵		+63 × 10⁵	

Table 7.7 *Terms in eastward momentum equation and tendency of eastward momentum* $(kg\, m^{-1}\, s^{-2})$.

Layer	Pressure gradient term $-\dfrac{\partial P}{\partial x}$	Coriolis term $+fV$	Remaining terms —	Tendency of momentum $\dfrac{\partial U}{\partial t}$	Ageostrophic wind $f(V - V_{geo})$
I	−3.13	−0.20	−0.05	−3.38	−3.33
II	−0.49	−0.70	+0.28	−0.91	−1.19
III	−0.94	+0.33	+0.20	−0.41	−0.61
IV	−1.29	+0.65	−0.07	−0.71	−0.64
V	−1.48	+0.62	+0.03	−0.83	−0.86
Mean abs. value	1.47	0.50	0.13	1.25	1.33

Table 7.8 *Terms in northward momentum equation and tendency of northward momentum* $(kg\, m^{-1}\, s^{-2})$.

Layer	Pressure gradient term $-\dfrac{\partial P}{\partial y}$	Coriolis term $-fU$	Remaining terms —	Tendency of momentum $\dfrac{\partial V}{\partial t}$	Ageostrophic wind $f(U - U_{geo})$
I	−2.41	+0.63	+0.22	−1.56	+1.78
II	+0.17	+1.64	−0.71	+1.10	−1.81
III	−0.23	+1.07	−0.20	+0.64	−0.84
IV	−0.77	+0.58	−0.01	−0.20	+0.19
V	−1.05	+1.24	+0.10	+0.29	−0.19
Mean abs. value	0.93	1.03	0.25	0.76	0.96

the lowest stratum would reach saturation within 18 hours. The tendencies in the other layers are also large and positive and would result in saturation of the entire troposphere within a few days, a highly improbable eventuality.

Finally, the stratospheric temperature prediction is clearly unreasonable. There are spuriously large vertical velocities associated with the strong divergence field. These give rise, through equation (5.21), to a large change in stratospheric temperature. Richardson's value, given in his Computing Form P_{XIV}, is $9.2 \times 10^{-4}\,K\,s^{-1}$, or 19.6 K in six hours. At this rate, the stratosphere would warm by 80 degrees in a day, an outlandish prediction.

7.4 The causes of the forecast failure

Richardson blamed the observed winds for the failure of his forecast. In his summary (Chapter 1 of *WPNP*) he wrote of how the forecast was spoilt by errors in the wind data. He continued: 'These errors appear to arise mainly from the irregular distribution of pilot balloon stations, and from their too small number' (*WPNP*, p. 2). Again, on page 187, he says 'This glaring error is . . . traced to errors in the representation of the initial winds.' In his discussion of convergence (*WPNP*, p. 212) he wrote: 'The striking errors in the 'forecast' . . . may be traced back to the large apparent convergence of wind.' He then asked whether the spurious convergence arose from the errors of balloon observations, or from the grid resolution being too large, or from the interpolation process from the observation points to the computation grid.

To investigate the calculation of divergence from observations, avoiding errors resulting from interpolation to a regular grid, Richardson computed the divergence directly from observations at three stations: Hamburg, Strasbourg and Vienna. These stations are approximately at the vertices of an equilateral triangle of side 700 km. He obtained a value of $3.05 \times 10^{-2}\,\mathrm{kg\,m^{-2}\,s^{-1}}$ for the total convergence in a column. This corresponds to a rise in pressure of about 60 hPa in six hours, or 10 Ber. He gave the following explanation: 'stations as far apart as 700 kilometres did not give an adequate representation of the wind in the lower layers'. This is a rather surprising conclusion. Elementary scale analysis shows that, for a given error in the observed wind, the error in the computed divergence actually increases as the spatial distance is reduced. Since the wind errors do not depend on the distance between stations, the error in computed divergence would actually be worse for stations more closely located. Richardson did not explicitly consider the compensation between influx and outflow that results in maintaining the divergence at a small value.

Shaw was fully aware of the importance of balance in the atmosphere and had a good understanding of it. In a preface to a Geophysical Memoir (Fairgrieve, 1913) on the closeness of observed winds to geostrophic balance, he wrote:

. . . the balance of [pressure] gradient and wind velocity is the state to which the atmosphere always tends, except in so far as it may be disturbed by the operation of new forces. If we suppose the balance once established, any disturbance by convection or otherwise must act in the free air by infinitesimal stages, and during every stage the tendency to restore the balance is continuously operative. Hence the transition from one set of conditions to another must be conducted by infinitesimal stages during which the disturbance of balance is infinitesimal.

This amounts to a clear, qualitative description of the process of geostrophic adjustment, which maintains the atmosphere in a state close to balance. Richardson joined the Met Office in 1913, the year this Memoir appeared and he must undoubtedly have studied it. He certainly was aware of the prevalance of quasi-geostrophic balance, but he also understood that the geostrophic wind was not useful for computing the divergence: '... for purposes not connected with finding the divergence of wind, the geostrophic hypothesis appears to serve as well in the stratosphere as elsewhere.' (*WPNP*, p. 146)

For his barotropic forecast (*WPNP*, Chapter 2), Richardson chose geostrophic initial winds. However, the implication of Eq. (4.8) (p. 68) that the pressure disturbances move westward led him to infer that the geostrophic wind is inadequate for the calculation of pressure changes. Purely geostrophic flow implies convergence for poleward flow and divergence for equatorward flow. On the basis of these results, he concluded that geostrophic winds would not serve adequately as initial conditions.

The initial winds tabulated and used by Richardson are wildly out of balance. There is huge disagreement between the pressure gradient and Coriolis terms in Tables 7.7 and 7.8. They should have values approximately equal in magnitude and opposite in sign. In the final columns of the tables, the components of the ageostrophic winds at each level are given (actually $f(U - U_{geo})$ and $f(V - V_{geo})$ are tabulated). The figures in the bottom rows show that the ageostrophic wind is comparable in magnitude to the total wind: the initial state is not remotely close to geostrophic balance.

Richardson placed great emphasis on the unrealistic divergence and little on the ageostrophic initial winds. He saw the solution as smoothing of the initial data to produce a realistic divergence field. However, it is insufficient to reduce the divergence to a small level: we saw in §4.6 that setting the divergence to zero at the initial time did not ensure that it remains small. Moreover, Richardson's analysis misses the point that what was needed was a *mutual adjustment*, which would ensure harmony between the mass and wind fields.

If the mass and wind are not in balance, large tendencies are inevitable. As a simple example, we consider the barotropic forecast again. If the initial wind field is shifted by 180°, the initial divergence is reversed in sign, so the the initial pressure tendency is unchanged in magnitude but of the opposite sign. However, the fields are now drastically out of balance: instead of cancelling, the pressure gradient and Coriolis terms reinforce each other. The initial momentum tendencies are now

$$\frac{\partial U}{\partial t} = f(V - V_{geo}) = 2fV \qquad \frac{\partial V}{\partial t} = -f(U - U_{geo}) = -2fU, \qquad (7.11)$$

suggesting a timescale $T = f^{-1} \approx 10^4$ s ≈ 3 hours, a small fraction of the synoptic timescale. Extrapolating these tendencies over an extended time interval results in a nonsensical forecast. In reality, such extreme initial conditions would engender massive gravity waves with wildly oscillating tendencies.

In summary, the spuriously large magnitude of the divergence of Richardson's initial winds, due to the absence of cancellation of terms arising from the compensation effects described in §7.2.3, made it inevitable that the initial pressure change would be catastrophic or, more correctly, 'anastrophic'. The large departure of his data from geostrophy meant that large initial momentum tendencies were unavoidable.

7.5 Max Margules and the 'impossibility' of forecasting

In a short paper published in the *Festschrift* to mark the sixtieth birthday of the renowned physicist Ludwig Boltzmann, Max Margules examined the relationship between the continuity equation and changes in surface pressure (Margules, 1904). A translation of this paper, together with a short introduction, has been published as a Historical Note by Met Éireann, the Irish Meteorological Service. Margules considered the possibility of predicting pressure changes by direct use of the mass conservation principle. He showed that, due to strong cancellation between terms, the calculation is very error-prone, and may give ridiculous results. Therefore, it is not possible, using the continuity equation alone, to derive a reliable estimate of synoptic-scale changes in pressure. Margules concluded that any attempt to forecast the weather was *immoral and damaging to the character of a meteorologist* (Fortak, 2001).

To make his forecast of the change in pressure, Richardson used the continuity equation, employing precisely the method that Margules had shown more than ten years earlier to be seriously problematical. As we have seen, the resulting prediction of pressure change was completely unrealistic. The question of what influence, if any, Margules' results had on Richardson's approach to forecasting was considered by Lynch (2003c). A copy of Margules' article was received and catalogued by the Met Office Library in March 1905. Thus, it was available for consultation by scientists such as Shaw and Dines. If Shaw, who was fully aware of Richardson's weather prediction project and indeed supported it strongly, knew of Margules' work, he would surely have alerted Richardson to its existence. There was ample opportunity for this between 1913, when Richardson was appointed Superintendent of Eskdalemuir Observatory, and 1916, when he resigned in order to work with the Friends Ambulance Unit in France.

Figure 7.1 Max Margules (1856–1920). Photograph from the archives of Zentralanstalt für Meteorologie und Geodynamik, Wien.

There is no reason to believe that Richardson was aware of Margules' paper; certainly, he makes no reference to it in his book. If he had been aware of Margules' results, he might well have decided not to proceed with his trial forecast, or sought a radically different approach (Platzman, 1967). Margules' results are summarised in Exner's *Dynamische Meteorologie*, published in 1917. It is possible that Richardson realised the significance of Margules' results when he read Exner's book. But, since he made no reference to the relevant section of Exner, it seems more likely that he simply overlooked it. At a later stage, Richardson did come to a realisation that his original method was unfeasible. In a note contained in the *Revision File*, inserted in the manuscript version of his book, he wrote 'Perhaps the most important change to be made in the second edition is that the <u>equation of continuity of mass must be eliminated</u>' (Richardson's underlining). Unfortunately, this note, which is reproduced in Platzman, 1967, is undated so we cannot say when Richardson reached this conclusion. He went on to speculate that the vertical component of vorticity might be a suitable prognostic variable. This was indeed a visionary anticipation of the use of the vorticity equation for the first successful numerical integration by Charney *et al.* in 1950.

Of course, we now know that Margules was unduly pessimistic. The continuity equation is an essential component of primitive equation models which are used in the majority of current computer weather prediction systems. These models support gravity wave solutions and, when changes in pressure are computed using the continuity equation, large tendencies can arise if the atmospheric conditions are far from balance. However, spuriously large tendencies are avoided in practice by an adjustment of the initial data to reduce gravity wave components to realistic amplitudes. This is the process of initialisation, which we discuss in detail in the next chapter.

8

Balance and initialisation

I still consider the elimination or dampening of noise to be the crucial
problem in weather analysis and prediction.   (*Hinkelmann, 1985*)

The spectrum of atmospheric motions is vast, encompassing phenomena having
periods ranging from seconds to millennia. The motions of interest for numerical
weather prediction have timescales greater than a day, but the mathematical equations describe a broader span of dynamical features than those of direct concern.
One of the long-standing problems in numerical weather prediction has been to
overcome the problems associated with high frequency motions. In this chapter
we will consider the history and development of methods that ensure that dynamic
balance of the initial data is achieved. We first discuss the phenomenon of balance
in the atmosphere. Then the valuable concept of the slow manifold is introduced.
We briefly review the principal methods for achieving balance in initial data. Next
we introduce a simple mechanical system, the *swinging spring*; the solutions to the
balance problem for this system have much wider applicability and are relevant to
the problems arising in general forecasting models. Finally, we discuss digital filter
initialisation, the technique that will be used to initialise Richardson's forecast.

8.1 Balance in the atmosphere

In Chapter 3 we examined the linear spectrum of atmospheric motions and saw
how the natural oscillations fall into two classes. The 'solutions of the first class'
are the rapidly-travelling high frequency gravity-inertia wave solutions, with phase
speeds of up to hundreds of metres per second and large divergence. The 'solutions
of the second class', which are the solutions of meteorological significance, are the
low frequency motions with phase speeds of the order of ten metres per second and
characteristic periods of a few days. The mass and wind fields of these solutions are
close to geostrophic balance and they are also called rotational or vortical modes,

137

since their vorticity is greater than their divergence; if divergence is ignored, these modes reduce to the Rossby–Haurwitz waves.

For typical conditions of large scale atmospheric flow (when the Rossby and Froude numbers are small) the two types of motion are clearly separated and interactions between them are weak. The high frequency gravity-inertia waves may be locally significant in the vicinity of steep orography, where there is strong thermal forcing or where very rapid changes are occurring, but overall they are of minor importance and may be regarded as undesirable noise.

A subtle and delicate state of balance exists in the atmosphere between the wind and pressure fields, ensuring that the fast gravity waves have much smaller amplitude than the slow rotational part of the flow. Observations show that the pressure and wind fields in regions not too near the equator are close to a state of geostrophic balance and the flow is quasi-nondivergent. The bulk of the energy is contained in the slow rotational motions and the amplitude of the high frequency components is small. The situation was described colourfully by Jule Charney, in a letter dated 12 February 1947 to Philip Thompson:

We might say that the atmosphere is a musical instrument on which one can play many tunes. High notes are sound waves, low notes are long [rotational] inertial waves, and nature is a musician more of the Beethoven than of the Chopin type. He much prefers the low notes and only occasionally plays arpeggios in the treble and then only with a light hand.

(in Thompson, 1983)

The prevalence of quasi-geostrophic balance is a perennial source of interest. It is a consequence of the forcing mechanisms and dominant modes of hydrodynamic instability and of the manner in which energy is dispersed and dissipated in the atmosphere. Observations show that the bulk of the energy in the troposphere is in rotational modes with advective timescales. The characteristic timescale of the external (solar) forcing is longer than the inertial timescale, so the forcing of gravity waves is weak. Coupling between the rotational and gravity wave components is also generally weak. Gravity waves interact nonlinearly to generate smaller spatial scales, which are heavily damped. They can propagate vertically, with amplitude growing until they reach breaking-point, releasing their energy in the stratosphere. This process is parameterised in many models. However, in the troposphere, gravity waves have little impact on the flow and may generally be ignored.

Classical reviews of geostrophic motion have been published by Phillips (1963) and Eliassen (1984). The gravity-inertia waves are instrumental in the process by which the balance is maintained, but the nature of the sources of energy ensures that the low frequency components predominate in the large scale flow. The atmospheric balance is subtle, and difficult to specify precisely. It is *delicate* in that minor perturbations may disrupt it but *robust* in that local imbalance tends to be rapidly

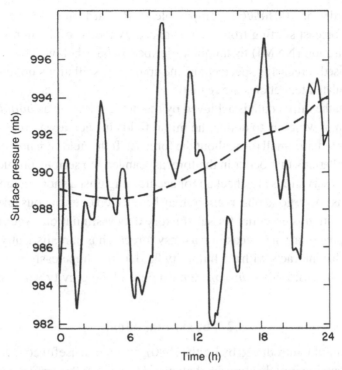

Figure 8.1 Surface pressure as a function of time for two 24-hour integrations of a primitive equation model. Solid line: unmodified analysis data. Dashed line: data initialised by the nonlinear normal mode technique. (From Williamson and Temperton (1981), © Amer. Met. Soc.)

removed through radiation of gravity-inertia waves in a spontaneous adjustment between the mass and wind fields. For a recent review of balanced flow, see McIntyre (2003).

When the primitive equations are used for numerical prediction the forecast may contain spurious large-amplitude high-frequency oscillations. These result from anomalously large gravity-inertia waves that occur because the balance between the mass and velocity fields is not reflected faithfully in the analysed fields. High frequency oscillations of large amplitude are engendered, and these may persist for a considerable time unless strong dissipative processes are incorporated in the forecast model. It was the presence of such imbalance in the initial fields that gave rise to the totally unrealistic pressure tendency of 145 hPa/6h obtained by Richardson. The noise problem is illustrated clearly in Fig. 8.1, from Williamson and Temperton (1981), which shows the evolution in time of the surface pressure at a particular location predicted by a primitive equation model. The solid line is for an integration from unmodified analysis data. There are oscillations with amplitudes up to 10 hPa and periods of only a few hours. The greatest instantaneous tendencies are of the order of 100 hPa/6h, comparable to Richardson's value.[1] The dashed line is for a forecast starting from data balanced by means of the nonlinear normal mode initialisation (NNMI) technique described in §8.3 below. The contrast with the uninitialised forecast is spectacular: the spurious oscillations no longer appear, and the calculated tendencies are realistic.

Balance in the initial data is achieved by the process known as *initialisation*, the principal aim of which is to define the initial fields in such a way that the gravity-inertia waves remain small throughout the forecast. If the fields are not initialised the spurious oscillations that occur in the forecast can lead to serious problems. In particular, new observations are checked for accuracy against a short-range forecast; if this forecast is noisy, good observations may be rejected or erroneous ones accepted. Thus, initialisation is essential for satisfactory data assimilation. Another problem occurs with precipitation forecasting: a noisy forecast has unrealistically large vertical velocity that interacts with the humidity field to give hopelessly inaccurate rainfall patterns. To avoid this *spin-up*, we must control the gravity wave oscillations.

8.2 The slow manifold

The slow manifold, introduced by Leith (1980), provides a useful conceptual framework for discussing initialisation and balance. The state of the atmosphere at a given time may be represented by a point $\mathbf{X}$ in a phase space $\mathcal{X}$. This state evolves in time according to an equation of the form

$$\dot{\mathbf{X}} + \mathcal{L}\mathbf{X} + \mathcal{N}(\mathbf{X}) = 0, \tag{8.1}$$

where $\mathcal{L}$ is a linear operator and $\mathcal{N}$ is a nonlinear function. The natural oscillations, or linear normal modes, of the atmosphere are obtained by spectral analysis of $\mathcal{L}$. We saw in Chapter 3 that they fall into two classes, so we can partition the state into orthogonal *slow* and *fast* components

$$\mathbf{X} = \mathbf{Y} + \mathbf{Z}.$$

The system (8.1) may then be separated into two sub-systems

$$\dot{\mathbf{Y}} + \mathcal{L}_Y \mathbf{Y} + \mathcal{N}_Y(\mathbf{Y}, \mathbf{Z}) = 0 \tag{8.2}$$

$$\dot{\mathbf{Z}} + \mathcal{L}_Z \mathbf{Z} + \mathcal{N}_Z(\mathbf{Y}, \mathbf{Z}) = 0. \tag{8.3}$$

The phase space $\mathcal{X}$ is the direct sum of two linear sub-spaces:

$$\mathcal{X} = \mathcal{Y} \oplus \mathcal{Z}.$$

---

[1] A gravity wave of amplitude $A = 10$ Pa and period $\tau = 100$ s would yield such a tendency: if $p = A \cos[kx - 2\pi t/\tau]$ then $|\partial p/\partial t|_{max} = 2\pi A/\tau \approx 0.6$ Pa s^{-1} or about 130 hPa/6h. For a phase speed $c = 300$ m s^{-1}, the wavelength is $L = 30$ km.

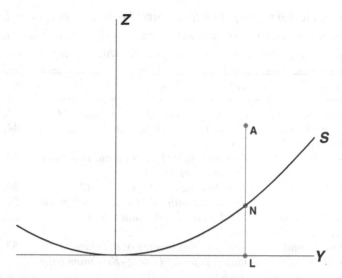

Figure 8.2 Schematic diagram showing the slow and fast linear sub-spaces $\mathcal{Y}$ and $\mathcal{Z}$ and the (nonlinear) slow manifold $\mathcal{S}$. Point $\mathbf{A} = (\mathbf{Y}, \mathbf{Z})$ represents the analysis. Point $\mathbf{L} = (\mathbf{Y}, \mathbf{0})$ is the result of linear normal mode initialisation and point $\mathbf{N} = (\mathbf{Y}, \mathbf{Z_B})$ is the result of nonlinear normal mode initialisation. (After Leith, 1980)

If nonlinearity is neglected, a point in either sub-space will remain therein as the flow evolves. This property defines $\mathcal{Y}$ and $\mathcal{Z}$ as *invariant sets* for linear flow. We call $\mathcal{Y}$ the linear slow sub-space. For the nonlinear system, $\mathcal{Y}$ is no longer invariant: interactions between slow waves generate fast oscillations through the term $\mathcal{N}_Z$ in (8.3). Is there a nonlinear counterpart of the linear slow subspace, that is invariant and contains only slowly varying flows? Leith proposed such a set, which may be thought of as a distortion of $\mathcal{Y}$, and he called it the *slow manifold*. We denote it by $\mathcal{S}$ and depict it schematically in Fig. 8.2. The slow and fast sub-spaces, which are multi-dimensional, are represented by the $\mathbf{Y}$ and $\mathbf{Z}$ axes and the manifold $\mathcal{S}$ by the curve $\mathbf{S}$. A key property of $\mathcal{S}$ is that it is of lower dimensionality than the phase space $\mathcal{X}$. In ideal circumstances, we may find a representation for $\mathcal{S}$ as a function of the slow variables. In that case, $\mathbf{Y}$ and $\mathbf{Z}$ are functionally related: $\mathbf{Z} = S(\mathbf{Y})$ on $\mathcal{S}$, and we may partition a state $\mathbf{X}$ in $\mathcal{S}$ into a slow part and a *balanced* fast part:

$$\text{For} \quad \mathbf{X} \in \mathcal{S}, \qquad \mathbf{X} = \mathbf{Y} + \mathbf{Z_B}, \qquad \text{where} \qquad \mathbf{Z_B} = S(\mathbf{Y}).$$

A flow whose state is initially in $\mathcal{S}$ should remain within it indefinitely ($\mathcal{S}$ is an invariant set) and should be free from high frequency gravity-inertia wave oscillations.

The slow manifold can be proven to exist for some simple equation systems. For more realistic models, and for the atmosphere, the question is more recondite. Table 8.1, which lists a series of papers addressing this question, indicates changing

Table 8.1 *A selection of papers in the* Journal of the Atmospheric Sciences *dealing with the existence or non-existence of a slow manifold in a simple five-component model, the Lorenz–Krishnamurthy model.*

Year	Author(s)	Title	Volume, Pages
1986	Lorenz, E. N.	*On the existence of a slow manifold.*	**43**, 1547–58.
1987	Lorenz, E. N. and Krishnamurthy, V.	*On the nonexistence of a slow manifold.*	**44**, 2940–50.
1991	Jacobs, S. J.	*Existence of a slow manifold in a model system of equations.*	**48**, 893–902.
1992	Lorenz, E. N.	*The slow manifold – what is it?*	**49**, 2449–51.
1994	Boyd, J. P.	*The slow manifold of a five-mode model.*	**51**, 1057–64.
1996	Fowler, A. C. and Kember, G.	*The Lorenz–Krishnamurthy slow manifold.*	**53**, 1433–7.
1996	Camassa, R. and Tin, S.-K.	*The global geometry of the slow manifold in the Lorenz–Krishnamurthy model.*	**53**, 3251–64.

views and some confusion, even for a relatively simple five-component model. It is now generally accepted that for realistic atmospheric models a slow manifold, in the strict mathematical sense, does not exist. Free unbalanced gravity-inertia waves propagating at their linear phase speeds are inevitably excited by nonlinear interactions of the balanced flow. Thus, the motion is not strictly 'slow'. However, the fast oscillations remain of small amplitude, their growth being limited by dissipative processes. In addition, the geometry of the set of balanced states is complex, with a Cantor-like structure, so it does not have the smoothness required of a manifold. Thus, to the question 'Does the slow manifold exist?', the purist must reply 'No!' The pragmatist may say 'Yes!', while recognising that, strictly speaking, it is not slow and it is not a manifold! Despite these mathematical subtleties, the concept of a slow manifold is of great value. The atmosphere remains *close to* a state of balance and deviations from it are localised and short-lived. We will show in §8.4 how the process of initialisation may be interpreted as a projection onto the slow manifold.

8.3 Techniques of initialisation

In order to avoid Richardson's difficulties, Charney, Fjørtoft and von Neumann (1950) made their first numerical forecast using the non-divergent barotropic vorticity equation. This has Rossby wave solutions but no high frequency gravity wave modes. More general systems of prediction equations with this property may be derived by modifying the primitive equations; this process is known as 'filtering'.

For several recent relevant papers, see Norbury and Roulstone (2002a,b). The basic filtered system is the set of quasi-geostrophic equations. These equations were used in operational forecasting for a number of years (see Chapter 10 for further discussion). However, filtered systems involve approximations that are not always valid, and this can result in poor forecasts, so means were sought to use more fundamental systems of equations.

Hinkelmann (1951) examined the question of initial conditions for the primitive equations. He recommended that the initial winds be derived using the geostrophic relation. He later successfully integrated the primitive equations, using a very short timestep (Hinkelmann, 1959). Forecasts made with the primitive equations were soon shown to be clearly superior to those using the quasi-geostrophic system. Hinkelmann had avoided the inaccuracies of filtered systems by returning to the primitive equations – but he had not solved the noise problem. Geostrophic initial winds are inadequate, as was shown already by Richardson in the 'introductory example' in Chapter 2 of his book. Charney (1955) proposed that a better estimate of the initial wind field could be obtained by using the nonlinear balance equation. This equation, which takes account of the curvature of the streamlines, is a diagnostic relationship between the pressure and wind fields. It implies that the wind is non-divergent.

The balance equation, which is derived by taking the divergence of the momentum equation and assuming non-divergent flow $\mathbf{V} = \mathbf{k} \times \nabla \psi$, may be written

$$\nabla^2 \Phi - \nabla \cdot f \nabla \psi - 2(\psi_{xx}\psi_{yy} - \psi_{xy}^2) = 0. \qquad (8.4)$$

For a given geopential Φ it is a nonlinear equation, known as a Monge–Ampère equation, for the stream function ψ. There are some technical difficulties with the solution of the balance equation, the most significant being the requirement that the data satisfy an *ellipticity criterion*. These problems were addressed in the late 1950s (Bolin, 1956; Árnason, 1958). It was later argued by Phillips (1960) that a further improvement of balance would result if the divergence of the initial field were set equal to that implied by quasi-geostrophic theory. This was obtained by deriving the vertical velocity from the omega equation and using the continuity equation. Each of these steps represented some progress, but the noise problem still remained essentially unsolved.

Sasaki (1958) formulated the initialisation process as a problem in variational calculus.[2] The modified fields were constrained to fit a balance condition as closely as possible whilst also remaining close to the original analysis; a geostrophic constraint was used in the 1958 paper. By a suitable choice of weighting functions, the wind

---

[a] The title of Sasaki's paper, 'An objective analysis based on the variational method', is somewhat misleading. The paper treats the problem of initialisation of analysed fields, not the analysis of data.

analysis could be given more emphasis in low latitudes and the height analysis in high latitudes. The method shared the shortcomings of the other methods discussed above in that geostrophic balance is inappropriate in equatorial regions. Although variational initialisation did not gain popularity, the variational method is at the core of modern data assimilation techniques. The method is particularly suited to assimilation of satellite data, and has been generalised from a single time technique to a process that covers a finite time-span, treating the temporal evolution of the atmosphere explicitly. This process, called four-dimensional variational assimilation (or 4D-Var, for short) is now in widespread use.

The methods considered so far may be described as *static initialisation* methods. Another approach, called *dynamic initialisation*, uses the forecast model itself to define the initial fields (Miyakoda and Moyer, 1968; Nitta and Hovermale, 1969). The dissipative processes in the model can damp out high frequency noise as the forecast proceeds. Moreover, numerical integration schemes having selective damping of high frequency components of the flow may be used. In dynamic initialisation, the model is integrated forward one time-step and then backward to the initial time, keeping the dissipation active all the while. This forward–backward cycle is repeated many times until we finally obtain fields, valid at the initial time, from which the high frequency components have been damped out. The forecast starting from these fields is noise-free. The procedure is expensive in computer time, and damps the meteorologically significant motions as well as the gravity waves, so it is no longer popular. However, the digital filtering technique, which will be discussed in §8.5, can be seen as a development of dynamical initialisation.

If the initial fields are separated into normal mode components, the gravity-inertia waves may be removed so as to leave only the slow rotational waves (Dickinson and Williamson, 1972). It had been hoped that this process of 'linear normal mode initialisation' would ensure a noise-free forecast. However, the results of the technique were disappointing. The noise soon reappeared: the slow components interact non-linearly in such a way as to generate gravity waves. The problem of noise remained.

After the establishment of the European Centre for Medium-Range Weather Forecasts (ECMWF), there was an active and urgent debate about finding a way of dealing with the problem. A satisfactory solution was found by Machenhauer (1977). He reasoned as follows: to control the growth of gravity waves, we set their initial rate of change to zero, in the hope that they will remain small throughout the forecast. The idea worked like a charm: forecasts starting from initial fields modified in this way are very smooth and the spurious gravity wave oscillations are almost completely removed (see Fig. 8.1). The method takes account of the nonlinear nature of the equations, and is referred to as 'nonlinear normal mode initialisation' (NNMI). Independently of Machenhauer, Baer (1977) developed a closely similar method that was based on more rigorous mathematical reasoning but was somewhat more difficult to apply (see also Baer and Tribbia, 1977). Thus, Machenhauer's method gained most popularity. The normal mode method was so successful that the long-standing problem of how to define initial fields was regarded as solved.

The application of the normal mode method requires the iterative solution of a nonlinear system of equations for the balanced fast modes. From (8.3) it is clear that $\dot{\mathbf{Z}} = \mathbf{0}$ implies

$$\mathcal{L}_Z \mathbf{Z} + \mathcal{N}_Z(\mathbf{Y}, \mathbf{Z}) = \mathbf{0}. \tag{8.5}$$

This must be solved for the balanced fast component $\mathbf{Z} = \mathbf{Z}_B$. Using the analysed coefficients to begin the process, we may define successive approximations using the Picard scheme:

$$\mathbf{Z}_{(k+1)} = -\mathcal{L}_Z^{-1} \mathcal{N}_Z(\mathbf{Y}, \mathbf{Z}_{(k)}).$$

This normally converges within a few iterations to a balanced value $\mathbf{Z}_B$ known as first-order balance. It is possible to define successively higher order balance relationships by requiring higher time derivatives of the fast variables to vanish. Zeroth-order balance requires $\mathbf{Z} = 0$, first-order balance requires $\dot{\mathbf{Z}} = 0$ and the n-th order balance condition is

$$\frac{d^n \mathbf{Z}}{dt^n} = 0.$$

We call this series of balance conditions the *Hinkelmann hierarchy*, as it was first proposed by Hinkelmann (1969). In principle, higher order initialisation schemes could be constructed in this way, but the diagnostic relationships for the fast variables become prohibitively complicated with increasing n. Lorenz (1980) introduced the concept of *super-balance*, defined by the condition

$$\lim_{n \to \infty} \frac{d^n \mathbf{Z}}{dt^n} = 0.$$

It was hoped that this condition would yield an exact characterisation of the slow manifold. As we have seen, more recent research has shown that such an ideal invariant manifold, devoid of freely propagating high frequency oscillations, does not exist. However, due to the quasi-balanced state of the atmosphere, the concept of a quasi-slow quasi-manifold remains of considerable practical value.

There are circumstances in which the normal mode method is difficult to apply. This is particularly the case for limited area models: for these models, the artificial boundaries introduce mathematical complexities such that the normal modes are in general unknown. In the 1980s, much effort was devoted to the problem of initialisation for these models, and several methods were devised (see Daley, 1991). More recently, a method using concepts from digital signal processing was developed, and is now in widespread use. This *digital filtering* initialisation technique will be described in §8.5, and its application to Richardson's data will be considered in Chapter 9.

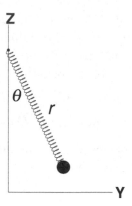

Figure 8.3 Diagram of the swinging spring. Polar co-ordinates (r, θ) about the point of suspension are used.

Daley (1991) devoted some six chapters of his book to various methods of initialisation, providing a comprehensive treatment of the subject. He discussed several other methods of initialisation, such as the bounded derivative method, the implicit normal mode method and the Laplace transform method, which we do not consider here. A more recent and much briefer review of initialisation is presented in Lynch (2003a). Daley extended Charney's musical analogy, writing 'we can think of inertia-gravity waves as meteorological noise that is not loud enough to drown out the symphony orchestra though it does detract from the performance. In the atmosphere, this noise is an inconvenience; in numerical models, it can be disastrous.' For Richardson, the noise was certainly disastrous but, as we will see, the disaster can be averted.

8.4 The swinging spring

The concepts of initialisation, filtering and the slow manifold can be vividly il-lustrated by considering the dynamics of a simple mechanical system governed by a set of ordinary differential equations (Lynch, 2002). The swinging spring, or elastic pendulum, is depicted in Fig. 8.3. It comprises a heavy bob suspended by a light elastic spring that can stretch but not bend. The bob is free to move in a vertical plane. The oscillations of this system are of two types, distinguished by their physical restoring mechanisms and their characteristic frequencies.

8.4.1 The linear normal modes

Let ℓ_0 be the unstretched length of the spring, ℓ its length at stable equilibrium, k its elasticity or stiffness and m the mass of the bob. At equilibrium the elastic restoring force is balanced by the weight:

$$k(\ell - \ell_0) = mg. \tag{8.6}$$

It is convenient to write the dynamic equations in Hamilton's canonical form.[3] If the total energy H is expressed in terms of the co-ordinates q_n and momenta p_n, these equations are

$$\frac{dq_n}{dt} = \frac{\partial H}{\partial p_n}, \qquad \frac{dp_n}{dt} = -\frac{\partial H}{\partial q_n}. \tag{8.7}$$

In the present case polar co-ordinates $q_r = r$ and $q_\theta = \theta$ are used, and the conjugate radial and angular momenta are $p_r = m\dot{r}$ and $p_\theta = mr^2\dot{\theta}$. The total energy is a sum of kinetic, elastic potential and gravitational potential energy:

$$H = \frac{1}{2m}\left(p_r^2 + \frac{p_\theta^2}{r^2}\right) + \tfrac{1}{2}k(r - \ell_0)^2 - mgr\cos\theta. \tag{8.8}$$

The canonical equations (8.7) may now be written explicitly

$$\dot{\theta} = p_\theta/mr^2 \tag{8.9}$$

$$\dot{p}_\theta = -mgr\sin\theta \tag{8.10}$$

$$\dot{r} = p_r/m \tag{8.11}$$

$$\dot{p}_r = p_\theta^2/mr^3 - k(r - \ell_0) + mg\cos\theta. \tag{8.12}$$

This system may be written symbolically in vector form

$$\dot{\mathbf{X}} + \mathbf{L}\mathbf{X} + \mathcal{N}(\mathbf{X}) = \mathbf{0}, \tag{8.13}$$

where $\mathbf{X} = (\theta, p_\theta, r, p_r)^{\mathrm{T}}$, $\mathbf{L}$ is the matrix of coefficients of the linear terms and $\mathcal{N}$ is a nonlinear vector function. This is formally identical to (8.1). The state vector $\mathbf{X}$ specifies a point in the four-dimensional phase space that defines the state of the system at any time; the motion is represented by the trajectory traced out by $\mathbf{X}$ as it moves through phase space.

Let us now suppose that the amplitude of the motion is small, so that $|r'| = |r - \ell| \ll \ell$ and $|\theta| \ll 1$. The equations may be linearised and written in matrix form

$$\frac{d}{dt}\begin{pmatrix} \theta \\ p_\theta \\ r' \\ p_r \end{pmatrix} = \begin{pmatrix} 0 & 1/m\ell^2 & 0 & 0 \\ -mg\ell & 0 & 0 & 0 \\ 0 & 0 & 0 & 1/m \\ 0 & 0 & -k & 0 \end{pmatrix}\begin{pmatrix} \theta \\ p_\theta \\ r' \\ p_r \end{pmatrix} \tag{8.14}$$

---

[3] If the formalism of Hamiltonian dynamics is unfamiliar, it may be noted that (8.10) and (8.12) are just Newton's equations of motion in polar co-ordinates with the forces resolved into azimuthal and radial components.

Since the matrix is block diagonal, this system splits into two sub-systems. The state vector $\mathbf{X}$ is partitioned into two sub-vectors

$$\mathbf{X} = \begin{pmatrix} \mathbf{Y} \\ \mathbf{Z} \end{pmatrix}, \qquad \text{where} \quad \mathbf{Y} = \begin{pmatrix} \theta \\ p_\theta \end{pmatrix} \quad \text{and} \quad \mathbf{Z} = \begin{pmatrix} r' \\ p_r \end{pmatrix},$$

and the linear dynamics of these components evolve independently:

$$\dot{\mathbf{Y}} = \begin{pmatrix} 0 & 1/m\ell^2 \\ -mg\ell & 0 \end{pmatrix} \mathbf{Y}, \qquad \dot{\mathbf{Z}} = \begin{pmatrix} 0 & 1/m \\ -k & 0 \end{pmatrix} \mathbf{Z}.$$

We call the motion described by $\mathbf{Y}$ the rotational component and that described by $\mathbf{Z}$ the elastic component. The rotational equations may be combined to yield

$$\ddot{\theta} + \omega_R^2 \theta = 0 \tag{8.15}$$

which is the equation for a simple harmonic oscillator or linear (anelastic) pendulum having oscillatory solutions with frequency $\omega_R = \sqrt{g/\ell}$. The remaining two equations yield

$$\ddot{r}' + \omega_E^2 r' = 0, \tag{8.16}$$

the equations for elastic oscillations with frequency $\omega_E = \sqrt{k/m}$. The solutions of (8.15) and (8.16) are called the linear normal modes. We assume the spring is very stiff, so that the rotational frequency is much smaller than the elastic:

$$\epsilon \equiv \left(\frac{\omega_R}{\omega_E} \right) = \sqrt{\frac{mg}{k\ell}} \ll 1. \tag{8.17}$$

Thus, the linear normal modes are clearly distinct: the rotational or swinging mode has low frequency (LF) and the elastic or springing mode has high frequency (HF). We consider the rotational mode as an analogue of the rotational modes in the atmosphere and the elastic mode as an analogue of the gravity-inertia modes.

8.4.2 Initialisation, the slow manifold and the slow equations

For small amplitude motions, for which the nonlinear terms are negligible, the LF and HF oscillations are completely independent of each other and evolve without interaction. We can suppress the HF component completely by setting its initial amplitude to zero:

$$\mathbf{Z} = \begin{pmatrix} r' \\ p_r \end{pmatrix} = \mathbf{0} \quad \text{at} \quad t = 0. \tag{8.18}$$

This is the procedure called *linear initialisation*. When the amplitude is large, nonlinear terms are no longer negligible. It is clear from the equations (8.9)–(8.12) that linear initialisation will not ensure permanent absence of HF motions: the nonlinear LF terms generate radial momentum. Machenhauer's (1977) proposal to minimise HF oscillations in such systems was to set the initial *tendency* of the HF components to zero:

$$\dot{\mathbf{Z}} = \begin{pmatrix} \dot{r} \\ \dot{p}_r \end{pmatrix} = 0 \quad \text{at} \quad t = 0. \tag{8.19}$$

This is the procedure called *nonlinear initialisation*. For the pendulum, we can deduce explicit expressions for the initial conditions by using (8.19) in (8.11) and (8.12):

$$r(0) = r_{\mathrm{B}} \equiv \frac{\ell[(1 - \epsilon^2) + \epsilon^2 \cos \theta]}{1 - \epsilon^2(\dot{\theta}/\omega_{\mathrm{R}})^2}, \qquad p_r(0) = 0. \tag{8.20}$$

Thus, given arbitrary initial conditions $\mathbf{X} = (\mathbf{Y}, \mathbf{Z}) = (\theta, p_\theta, r, p_r)^{\mathrm{T}}$, we replace the HF component $\mathbf{Z} = (r, p_r)^{\mathrm{T}}$ by $\mathbf{Z}_{\mathrm{B}} = (r_{\mathrm{B}}, 0)^{\mathrm{T}}$. The rotational component $\mathbf{Y} = (\theta, p_\theta)^{\mathrm{T}}$ remains unchanged. If, for simplicity, we assume that the angular momentum p_θ vanishes at $t = 0$, the condition $r = r_{\mathrm{B}}$ defines a curve in the (r, θ)-plane:

$$r = \ell[(1 - \epsilon^2) + \epsilon^2 \cos \theta]. \tag{8.21}$$

This is one of the classical 'special curves', called the limaçon of Pascal, named after Étienne, father of Blaise Pascal (Wells, 1991).

The linear initialisation condition $\mathbf{Z} = 0$ defines a two-dimensional sub-space of phase-space, the plane surface $\mathcal{Y}$ through the origin given by $r = p_r = 0$. However, for nonlinear motion a point initially in this plane will not remain therein. The nonlinear initialisation condition $\dot{\mathbf{Z}} = 0$ defines a two-dimensional nonlinear sub-set, the surface $\mathcal{S}_1$ given by $r = r_{\mathrm{B}}$, $p_r = 0$. This surface is a first-order approximation to the slow manifold. While a point initially on $\mathcal{S}_1$ is not guaranteed to remain on it, we find that points initially close to $\mathcal{S}_1$ remain close to it for all time: $\mathcal{S}_1$ acts like a guiding centre for the motion. Nonlinear initialisation ensures that the initial point is on $\mathcal{S}_1$. It can be represented as a surface in three-space, since $p_r = 0$ (for a plot of this surface, see Lynch (2002), Fig. 3). The cross-section through the plane $p_\theta = 0$ is the limaçon $r = \ell_0(1 + \epsilon^2 \cos \theta)$. Limaçon, meaning snail, is an apposite name for a curve circumscribing the slow manifold. For small amplitude motions (θ and $\dot{\theta}/\omega_{\mathrm{R}}$ small), the surface $\mathcal{S}_1$ may be approximated by a quadratic surface:

$$\hat{r}_{\mathrm{B}} \equiv \ell - r_{\mathrm{B}} = \epsilon^2 \ell \left[\tfrac{1}{2}\theta^2 - \left(\frac{\dot{\theta}}{\omega_{\mathrm{R}}} \right)^2 \right]. \tag{8.22}$$

This surface, a hyperbolic paraboloid, has a saddle point at the origin. The cross-section for $\dot{\theta} = 0$ is a parabola, similar to the schematic diagram in Fig. 8.2 (p. 141).

The point $\mathbf{A}$ in the figure represents the analysed initial state. Point $\mathbf{L}$, which results from (8.18), is a projection onto the linear subspace $\mathcal{Y}$. Point $\mathbf{N}$ is the result of nonlinear initialisation (8.19). Thus we see that the process of nonlinear initialisation is a projection, parallel to $\mathcal{Z}$, onto the manifold $\mathcal{S}_1$, leaving the slow component $\mathbf{Y}$ unchanged.

We can artificially force the motion to evolve on the slow manifold $\mathcal{S}_1$ by modifying, or *filtering*, the dynamical equations. We replace the prognostic equations for the radial motion by the diagnostic equations $\dot{\mathbf{Z}} = \mathbf{0}$ or $\mathbf{Z} = \mathbf{Z}_B$. This yields the slow equations (Lynch, 1989). The system becomes

$$\dot{\theta} = p_\theta/mr^2 \tag{8.23}$$

$$\dot{p}_\theta = -mgr \sin \theta \tag{8.24}$$

$$0 = p_r/m \tag{8.25}$$

$$0 = p_\theta^2/mr^3 - k(r - \ell_0) + mg \cos \theta. \tag{8.26}$$

The slow equations have linear normal mode solutions corresponding to the rotational motions, with frequency $\omega_R = \sqrt{g/\ell}$. There are no HF normal modes; they have been filtered out by the condition $\dot{\mathbf{Z}} = \mathbf{0}$. The slow equations describe dynamics on the manifold $\mathcal{S}_1$. It must be stressed that the solutions of this system are not an exact representation of the dynamics of the swinging spring, but they are a close approximation to the full dynamics provided the amplitude of the HF component remains small, so that the trajectory for the full system remains close to $\mathcal{S}_1$.

8.4.3 Numerical solutions

We will illustrate the effect of initialisation on the behaviour of the swinging spring by presenting results of numerical integrations of the governing equations with two sets of initial conditions. The parameter values for both runs are $m = 1 \, \mathrm{kg}$, $\ell = 1 \, \mathrm{m}$, $g = \pi^2 \, \mathrm{m \, s^{-2}}$ and $k = (10\pi)^2 \, \mathrm{kg \, s^{-2}}$ so that $\epsilon = 10^{-1}$. The linear rotational mode has frequency $\omega_R = \pi$ (period 2 seconds) and the frequency of the elastic mode is ten times greater, $\omega_E = 10\pi$.

The system (8.9)–(8.12) is solved by the Bulirsch–Stoer method (Press *et al.*, 1992, sub-routine BSSTEP) which is a modern implementation of Richardson's deferred approach to the limit (Richardson, 1927). The first set of initial conditions are

$$\mathbf{X}_0 = \left(\theta(0), p_\theta(0), r(0), p_r(0)\right)^{\mathrm{T}} = \left(1, 0, 1, 0\right)^{\mathrm{T}}.$$

These satisfy the condition of linear normal mode initialisation (LNMI), namely $\mathbf{Z} = (r', p_r)^{\mathrm{T}} = \mathbf{0}$ at $t = 0$. The second set of initial conditions are derived by means of the condition (8.20) for nonlinear normal mode initialisation (NNMI). This gives

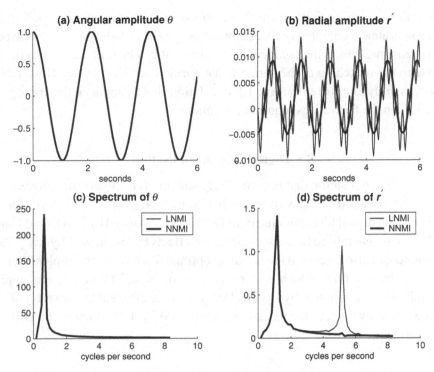

Figure 8.4 Numerical integration of swinging spring equations. Thin lines are for the integration from linearly initialised conditions (LNMI) and thick lines are for nonlinearly initialised conditions (NNMI). Panel a: evolution of the slow variable θ with time. Panel b: evolution of the fast variable r'. Panel c: spectrum of the slow component θ. Panel d: spectrum of the fast component r'.

a value $r(0) = r_B \approx 0.99540$. The results are presented in Fig. 8.4; the integration time is six seconds in both cases. The variation of θ with time is shown in panel (a). The two curves for this slow variable are so close as to be indistinguishable on the plot. The fast variable, the radial amplitude $r' = r - \ell$, appears in panel (b) of Fig. 8.4. The HF component is clearly visible for the linearly initialised run (the thin line); note that $\omega_E = 10\pi$ means five cycles per second. The amplitude is small – about 5 mm – thanks to the linear initialisation, but still much larger than for the NNMI run: the HF variation is not apparent for nonlinear initialisation (thick line). Both runs also have variations with a period of about one second: the nonlinear centrifugal force stretches the pendulum when the angular momentum is large; this happens twice in each rotational cycle or about once per second.

The spectra of the slow and fast components are shown in the lower panels of Fig. 8.4. The period of the slow motion is two seconds. This is confirmed by a large peak in the spectrum of θ at 0.5 Hz (cycles per second) seen in panel (c). For the fast variable r' (panel (d)), both runs have a large peak at 1 Hz. The high frequency

noise in the linearly initialised run is confirmed by the pronounced peak at 5 Hz.
This peak is almost completely absent for the nonlinear initialisation; only the peak
at 1 Hz remains. This is the *balanced fast motion*. It can be understood physically:
it is due to the stretching of the spring twice during each slow cycle. Thus, NNMI
has successfully eliminated the free HF oscillations but maintained the appropriate
projection on the fast modes required for balance.

8.5 Digital filter initialisation

A method of initialisation that is particularly suitable for limited area models, and
which has several advantages over alternative methods, will now be considered.
The method of digital filter initialisation (DFI) was introduced by Lynch and Huang
(1992). It was generalised to allow for diabatic effects by Huang and Lynch (1993).
The latter paper also discussed the use of an optimal filter. A much simpler filter, the
Dolph–Chebyshev filter, which is a special case of the optimal filter, was applied
to the initialisation problem by Lynch (1997). A more efficient formulation of DFI
was presented by Lynch, Giard and Ivanovici (1997). For a review of DFI, see
Lynch (2003b).

8.5.1 The Dolph–Chebyshev filter

Consider a sequence of values $\{\cdots, x_{-2}, x_{-1}, x_0, x_1, x_2, \cdots\}$ of a quantity whose
variation has both low and high frequency components. For example, x_n could
be the value of surface pressure at a particular grid point at time $t = n\Delta t$. We
call the sequence $\{x_n\}$ the input signal. The shortest period component that can
be represented with a time step Δt is $\tau_N = 2\Delta t$, corresponding to a maximum
frequency, the so-called Nyquist frequency, $\omega_N = \pi/\Delta t$. We define the digital
frequency $\theta = \omega\Delta t$ and note that it falls in the range $[-\pi, +\pi]$.

To separate the low frequency and high frequency components of the input signal,
we use a *finite impulse response* (FIR) filter, or non-recursive filter, defined by

$$y_n = \sum_{k=-N}^{N} h_k x_{n-k}. \tag{8.27}$$

Thus, each term y_n in the *output* signal $\{y_n\}$ is a weighted sum of input values
at times surounding time $n\Delta t$. The summation in (8.27) is formally identical to a
truncated, discrete convolution. (For a general treatment of digital filtering, see e.g.
Hamming (1989) or Oppenheim and Schafer (1989)).

The transfer function $H(\theta)$ of an FIR filter is defined as the function by which
a pure sinusoidal oscillation is multiplied when passed through the filter. If $x_n =$

$\exp(in\theta)$, one may write $y_n = H(\theta)x_n$, and $H(\theta)$ is easily calculated by substituting x_n in (8.27):

$$H(\theta) = \sum_{k=-N}^{N} h_k \exp(-ik\theta). \tag{8.28}$$

For symmetric coefficients $h_k = h_{-k}$ this is real, implying that the phase is not altered by the filter:

$$H(\theta) = \left[h_0 + 2\sum_{k=1}^{N} h_k \cos k\theta \right]. \tag{8.29}$$

An *ideal* low-pass filter with a cut-off frequency θ_c has the response

$$H_{\text{ideal}}(\theta) = \begin{cases} 1, & |\theta| \le |\theta_c|; \\ 0, & |\theta| > |\theta_c|, \end{cases} \tag{8.30}$$

This response cannot be realised by an FIR filter. The problem of filter design is to define the filter weights h_n so as to achieve a good approximation to the desired frequency response.

We now consider a particularly simple filter, having explicit expressions for its impulse response coefficients h_n. The details of the Dolph–Chebyshev filter are presented in Lynch (1997). We will limit the present discussion to the definition and principal properties of the filter. The function to be described is constructed using Chebyshev polynomials, defined by the equations

$$T_n(x) = \begin{cases} \cos(n \cos^{-1} x), & \text{if } |x| \le 1; \\ \cosh(n \cosh^{-1} x), & \text{if } |x| > 1. \end{cases}$$

Clearly, $T_0(x) = 1$ and $T_1(x) = x$. From the definition, the following recurrence relation follows immediately:

$$T_n(x) = 2x T_{n-1}(x) - T_{n-2}(x), \quad n \ge 2.$$

The main relevant properties of these polynomials are given in Lynch (1997).

Let us consider the function defined in the frequency domain by

$$H(\theta) = \frac{T_{2M}(x_0 \cos(\theta/2))}{T_{2M}(x_0)}$$

where $x_0 > 1$, and x_0 and M will be chosen to determine the desired characteristics of the filter. Let θ_s be such that $x_0 \cos(\theta_s/2) = 1$. As θ varies from 0 to θ_s, $H(\theta)$ falls from 1 to $r = 1/T_{2M}(x_0)$, where $r < 1$, the *ripple ratio*, is defined by

$$r = \frac{\text{side-lobe amplitude}}{\text{main-lobe amplitude}}$$

and is a measure of the maximum amplitude in the stop-band $[\theta_s, \pi]$, For $\theta_s \leq \theta \leq \pi$, $H(\theta)$ oscillates in the range $[-r, +r]$. The form of $H(\theta)$ is that of a low-pass filter with a cut-off at $\theta = \theta_s$. By means of the definition of $T_n(x)$ and basic trigonometric identities, $H(\theta)$ can be written as a *finite expansion*

$$H(\theta) = \sum_{n=-M}^{+M} h_n \exp(-in\theta).$$

This is formally identical to (8.28) above. The coefficients $\{h_n\}$ may be evaluated from the inverse Fourier transform

$$h_n = \frac{1}{N} \left[1 + 2r \sum_{m=1}^{M} T_{2M} \left(x_0 \cos \frac{\theta_m}{2} \right) \cos m\theta_n \right],$$

where $|n| \leq M$, $N = 2M + 1$ and $\theta_m = 2\pi m/N$ (Antoniou, 1993). Since $H(\theta)$ is real and even, h_n is also real and $h_{-n} = h_n$. The weights $\{h_n : -M \leq n \leq +M\}$ define the Dolph–Chebyshev or, for short, Dolph filter. The Dolph filter has minimum ripple-ratio for given main-lobe width and filter order. The filter order $N = 2M + 1$ is determined by the time step Δt and forecast span T_S. The desired frequency cut-off is specified by choosing a value for the cut-off period, τ_s. Then $\theta_s = 2\pi \Delta t/\tau_s$ and the parameters x_0 and r are given by

$$\frac{1}{x_0} = \cos \frac{\theta_s}{2}, \qquad \frac{1}{r} = \cosh \left(2M \cosh^{-1} x_0 \right).$$

Let us suppose components with period less than three hours are to be eliminated ($\tau_s = 3\,\mathrm{h}$) and the time step is $\Delta t = 300\,\mathrm{s}$. Then $\theta_s = 2\pi \Delta t/\tau_s \approx 0.1745$. It is found that a filter of order $N = 37$, or span $T = 2M\Delta t = 3\,\mathrm{h}$, attenuates high frequency components by more than $20\,\mathrm{dB}$ (the attenuation in the stop-band is $\delta = 20 \log_{10} |H(\theta)|$, measured in decibels). This level of damping implies that the amplitudes of high-frequency components are reduced by at least 90 per cent and their energy by at least 99 per cent, which is found to be adequate in practice. The frequency response of the Dolph–Chebyshev filter with these parameters is shown in Fig. 8.5. All the lobes in the stop-band have the same amplitude, equal to the ripple ratio $r = 0.0859$, giving rise to the name equiripple filter for filters of this type. In Lynch (1997, Appendix) it is proved that the Dolph window is an *optimal* filter whose pass-band edge, θ_p, is the solution of the equation $H(\theta) = 1 - r$.

8.5.2 *Application to initialisation*

We now describe how a digital filter may be used to initialise the analysis for a numerical forecast. The uninitialised fields of surface pressure, temperature and winds are represented by a vector $\mathbf{X}_0$. From these initial conditions, the model is

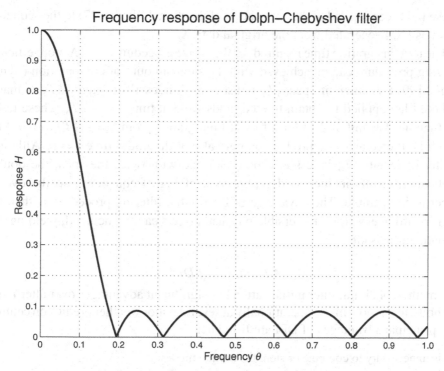

Figure 8.5 Frequency response of a Dolph–Chebyshev filter with parameters $M = 18$, $N = 37$, $\Delta t = 300$ s and $\tau_S = 3$ h. The digital stop-band edge is at $\theta_s = 0.1745$ and the ripple ratio is $r = 0.0859$.

integrated forward for a time $\frac{1}{2}T_S$, half the filter span. Running sums of the form

$$\mathbf{Y}_F = \frac{1}{2}h_0\mathbf{X}_0 + \sum_{n=1}^{N} h_{-n}\mathbf{X}_n, \tag{8.31}$$

where $\mathbf{X}_n = \mathbf{X}(n\Delta t)$, are accumulated. The summations are done for each variable at each gridpoint and on each model level. These are stored at the end of the short forecast. The original fields $\mathbf{X}_0$ are then used to make a three hour 'hindcast' from $t = 0$ to $t = -\frac{1}{2}T_S$, during which running sums of the form

$$\mathbf{Y}_B = \frac{1}{2}h_0\mathbf{X}_0 + \sum_{n=-1}^{-N} h_{-n}\mathbf{X}_n \tag{8.32}$$

are accumulated for each field, and stored as before. The two results are then combined to form the required summations:

$$\mathbf{Y}_0 = \mathbf{Y}_F + \mathbf{Y}_B = \sum_{n=-N}^{N} h_{-n}\mathbf{X}_n. \tag{8.33}$$

These fields correspond to the application of the digital filter (8.27) to the sequence of states $\{\mathbf{X}_n\}$ generated from the original data $\mathbf{X}_0$.

The total integration time required for the above procedure is T_S. A more efficient filtering procedure can be achieved with the same amount of computation (Lynch *et al.*, 1997). A filter of span $\frac{1}{2}T_S$ is used. A 'hindcast' of length $\frac{1}{2}T_S$ is made, and the filter applied to obtain filtered fields valid at time $t = -\frac{1}{4}T_S$. These fields are then used as initial data for a forward integration from $t = -\frac{1}{4}T_S$ to $t = \frac{1}{4}T_S$, which is filtered again to produce output valid at the initial time $t = 0$. Although a filter of length $\frac{1}{2}T_S$ is less effective than one with twice the span, the double application results in increased suppression of high frequencies, so the overall effect is comparable. The advantage of the double filtering procedure is that, for the forward integration, irreversible physical effects can be included, yielding full diabatic initialisation.

Advantages of DFI

The method of digital filter initialisation has significant advantages over alternative methods, and is now in use operationally at several major weather prediction centres. The principal advantages of DFI are listed here:

- it is unnecessary to compute or store the normal modes;
- it is unnecessary to separate the vertical modes;
- there is complete compatibility with the model discretisation;
- the method is applicable to exotic grids on arbitrary domains;
- there is no iterative numerical procedure that can diverge;
- the method is simple to implement and to maintain;
- filtering can be applied to all prognostic model variables;
- spin-up of moisture variables is greatly reduced;
- the method is immediately applicable to non-hydrostatic models.

For models that employ discretisations with stretched or irregular horizontal grids or exotic vertical co-ordinates, the normal mode structure may be difficult to obtain. DFI is applicable to such models without difficulty. For example, the method is used in the rapid update cycle (RUC) model of the National Centers for Environmental Prediction (NCEP) (Benjamin *et al.*, 2004), which has a hybrid σ-θ vertical co-ordinate, and in the GME model of the German Weather Service (Majewski *et al.*, 2002) which uses an icosahedral-hexagonal horizontal grid. Additional prognostic model variables, such as cloud water, rain water, turbulent kinetic energy, etc., are processed in the same way as the standard mass and wind variables; thus, DFI produces filtered initial fields for these variables that are compatible with the basic dynamical fields. In particular, DFI filters the additional prognostic variables in non-hydrostatic models in a manner identical to the basic variables. The DFI method is thus immediately applicable to such models (Bubnová *et al.*, 1995; Doms and Schättler, 1998). Finally, digital filters can be used to construct balance constraints suitable for use in variational data assimilation. This has been done at several operational forecasting centres.

9

Smoothing the forecast

The scheme of numerical forecasting has been developed so far that it is
reasonable to expect that when the smoothing of Ch. 10 has been arranged,
it may give forecasts agreeing with the actual smoothed weather.

(*WPNP*, p. 217)

In this chapter we present the results of a computer model reconstruction of
Richardson's forecast. His results are reproduced faithfully by the model. When
the initial fields are processed by a digital filter, the magnitudes of the tendencies
are drastically reduced, and the predicted changes are realistic. The effects of the
filter are illustrated by comparison of the original and filtered fields. Although the
changes made by the filtering procedure are generally small, they have a major
effect on the character of the forecast. To extend the prediction to 24 hours, additional spatial smoothing is required during the integration. This is consistent with
general experience in numerical prediction models: artificial numerical damping is
usually required to ensure stability of the integration.

9.1 Reconstruction of the forecast

A computer program has been written to repeat and extend Richardson's forecast.
It is a realisation of the algorithm presented in detail in §5.6 above. The forecast
starts from the same initial values as Richardson used, so that the calculated initial
changes could be compared directly with the values in *WPNP*. It will be seen that
the computer model produces results consistent with those obtained manually by
Richardson. In particular, the 'glaring error' in the surface pressure tendency is
reproduced almost exactly by the model.

Richardson computed the initial tendencies by evaluating the right-hand sides
of equations of the form

$$\frac{\partial Q}{\partial t} = R \tag{9.1}$$

Table 9.1 *Six-hour changes in pressure (hPa per 6
hours) obtained by Richardson (column 2) and
reconstructed using the computer program (column 3).*

	Pressure changes	
Level	Richardson	Reconstruction
1	48.3	48.5
2	77.0	76.7
3	103.2	102.1
4	126.5	124.5
S	145.1	145.4

for each prognostic variable Q. Using the leapfrog method to approximate the
time derivative, the change in Q over the time interval between $t = (n - 1)\Delta t$ and
$t = (n + 1)\Delta t$ is:

$$\Delta Q = Q^{n+1} - Q^{n-1} = 2\Delta t \times R^n. \tag{9.2}$$

It is important to emphasise that the tendency R^n is *independent of the time step*
used to discretise the equation. The time step between successive calculations of
R specified by Richardson was $\Delta t = 3$ h. This large value is frequently, and quite
incorrectly, given as the reason for the unrealistic tendency values he obtained.
The time step is hugely important in determining the numerical stability of the
integration, but has no bearing on the values of the initial tendencies. The changes
given in *WPNP* are over a six-hour period *centred* at the initial time 0700 UTC
20 May 1910. Thus, Richardson expressed his forecast results as changes between
0400 UTC and 1000 UTC on 20 May 1910. However, this was symbolic: he could
not explicitly compute Q^{n+1} for 1000 UTC from (9.2) as he did not have the values
Q^{n-1} for 0400 UTC. Thus, his 'forecast' was confined to the expression of the
computed changes only.

    We first consider the changes of the pressure at each of the four layer inter-
faces and at the Earth's surface for the central P-point. The values obtained by
Richardson are given in Table 9.1 (they are in the column marked Richardson). The
corresponding changes produced by the numerical model are also given, in the column marked Reconstruction. The units of pressure change are hPa/6h. It is evident
that the changes computed by the model are in close agreement with Richardson's
calculations; the reasons for the small discrepancies are discussed below.

    The changes of momentum calculated by Richardson and those computed with
the model are given in Table 9.2. The correspondence is not as close as for the
pressure changes, but there is broad agreement between the two sets of forecast

Table 9.2 *Six-hour changes in momentum components*
$(10^3 \ kg \ m^{-1} s^{-1} \ per \ 6 \ hours)$.

Values obtained by Richardson (columns 2 and 4) and values reconstructed using
the computer program (columns 3 and 5) are given.

Layer	Eastward momentum changes		Northward momentum changes	
	Richardson	Reconstruction	Richardson	Reconstruction
I	−73.0	−71.8	−33.7	−39.7
II	−19.6	−19.9	+23.8	+29.0
III	−8.9	−10.3	+13.8	+15.9
IV	−15.3	−13.7	−4.3	−4.1
V	−17.9	−22.5	+6.3	+7.1

changes. Richardson also presented his results for the change in water content in
each layer. Since all hydrological processes have been ignored in the computer
program written to repeat the forecast, these will not be considered further. The
remaining prognostic variable is the temperature of the uppermost layer. The change
tabulated by Richardson was $\Delta T_1 = 19.6°$. This was calculated using a complicated
equation; a much simplified version of the equation yielded a forecast change of
$\Delta T_1 = 19.9°$. The computer model used the simpler equation (5.20) and gave a
forecast change of $\Delta T_1 = 19.6°$, in close agreement with Richardson.

The reconstructed forecast is in good agreement with Richardson's values. Nevertheless, there are differences that require explanation. There are several reasons
why the results are not identical, of which the following are perhaps the most
important:

- Richardson's calculations were correct to three or four significant digits. The computer
  has much greater accuracy than this.
- Richardson made a number of minor arithmetical errors that were undetected despite his
  duplication of all computations.
- The reconstruction model omitted numerous physical processes such as moist physics,
  radiation and turbulence.
- The reconstruction model assumed a constant value for the acceleration of gravity, and a
  number of other minor simplifications were made.
- In the momentum equations, Richardson used w at the adjacent P-point 'for illustration' (*WPNP*, p. 186). In the reconstruction model, w is obtained by interpolation from
  surrounding P-points.
- Richardson's vertical interpolation for surface pressure appears to be error-prone.

The above results confirm the essential correctness of Richardson's calculations.
Such manual calculations, involving extensive numerical computation, are no-

toriously prone to error. Although a number of minor slips were detected in Richardson's calculations, they did not have any significant effect on his results. We may conclude that his unrealistic results were not due to arithmetical blunders, but had a deeper cause.

9.2 Richardson's five smoothing methods

Richardson devoted a short chapter of *WPNP* to the problem of smoothing the initial data for the forecast. Although it is only three pages long, it contains some crucially important ideas. It also illustrates the extent of Richardson's understanding of the causes of his forecast failure. He opens with a discussion of the impossibility of representing atmospheric phenomena having scales smaller than the computational grid, and of the need to represent such phenomena by introducing appropriate eddy diffusion terms in the equations. He remarks that 'So far meteorologists do not appear to have attended to eddy-diffusivities of this kind.' He then presents five ways in which smoothing may be achieved, and we will consider them in turn, under the headings used by Richardson.

A. SPACE MEANS. If the density of observation stations were such that several independent observations were available for a given grid cell, some sort of spatial average of the observed values would be an appropriate quantity to use as initial data. However, due to the scarcity of observations, this was not seen to be a practicable procedure. (The method is used today for some satellite data, and is called *super-obbing*.)

B. TIME MEANS. A more feasible procedure would be to take a series of observed values from each observation station, and use the average over the time span as the initial datum. This method of smoothing was seen by Richardson as the most practical of the five methods considered. The digital filtering initialisation technique is conceptually similar to averaging time-series of observations, with two important distinctions: the values to be filtered are generated by the forecast model, and a more selective filter is employed.

C. POTENTIAL FUNCTION. Richardson noted that the 'irregularity in the observations which has forced itself on our attention' is the large value of momentum divergence. The momentum vector $\mathbf{V} = (U, V)$ may be resolved into divergent and rotational components

$$\mathbf{V} = \nabla \chi + \mathbf{k} \times \nabla \psi.$$

For a global domain, this partitioning is unique. The momentum divergence and vorticity are given by

$$\delta = \nabla^2 \chi, \qquad \zeta = \nabla^2 \psi.$$

To reduce the divergence, we define a new momentum field

$$\hat{\mathbf{V}} = \mathbf{V} - \alpha_\delta \nabla \chi, \tag{9.3}$$

where $\alpha_\delta \in [0, 1]$ is a constant. The vorticity is unchanged by this transformation, but $\nabla \cdot \hat{\mathbf{V}}$ is a factor $(1 - \alpha_\delta)$ smaller than $\nabla \cdot \mathbf{V}$. For $\alpha_\delta = 1$ the modified momentum is divergence-free. Alternatively, Richardson proposed that α_δ might be chosen as a function of the observed barometric tendency. However, a spatially varying coefficient α_δ is itself a source of spurious divergence, as is easily seen by differentiating (9.3).

D. STREAM FUNCTION. Richardson noted that the curl of momentum was 'almost certainly irregular and in need of preliminary smoothing if we are to avoid awkward consequences in the application of the dynamical equations.' He proposed a transformation of the form

$$\hat{\mathbf{V}} = \mathbf{V} - \alpha_\zeta \mathbf{k} \times \nabla \psi, \tag{9.4}$$

with $\alpha_\zeta \in [0, 1]$, that would 'remove or diminish the irregular curl', without affecting the divergence.

E. SMOOTHING DURING THE FORECAST. The final method proposed by Richardson was the introduction of fictitious viscosity into the dynamical equations. This would be kept active until the irregularities in the flow had been sufficiently smoothed. He noted that the smoothing would act on all irregularities 'whether waves of compression or whirls'. That is, both the divergent and rotational components of the flow would be damped.

In his discussion of smoothing, Richardson treated damping of divergence and of vorticity on an equal footing. Although he identified the large value of momentum divergence as the primary cause of his forecast failure, his methods C and D would reduce or annihilate divergence and vorticity respectively. In the extreme case $\alpha_\delta = \alpha_\zeta = 1$, the momentum would be set identically to zero. Moreover, his artificial viscosity (method E) was intended to act on both the divergence and vorticity fields. Such an effect might be achieved by adding a diffusion term to the momentum equation,

$$\frac{\partial \mathbf{V}}{\partial t} = \mathbf{F} + \kappa \nabla^2 \mathbf{V},$$

where $\mathbf{F}$ represents the usual terms and κ is a diffusion coefficient. This acts selectively on the smaller spatial scales, damping both the δ and ζ fields. However, damping that acts exclusively on the divergent component of the flow may be achieved by adding a term as follows:

$$\frac{\partial \mathbf{V}}{\partial t} = \mathbf{F} + \kappa \nabla \delta.$$

This has no effect on the vorticity field (as is clear by taking the curl of the equation). Such divergence damping has been used in atmospheric models (Sadourny, 1975). It is highly unlikely that the damping of vorticity, Richardson's Method D, would yield satisfactory results.

We must observe that nowhere in Richardson's discussion is the need for a *mutual adjustment* between the mass and wind fields mentioned. If the initial fields are close to balance, imposition of damping on the momentum field will actually disimprove the situation by driving the atmospheric state away from balance. It is the mutual adjustment of the fields that is the task of initialisation.

9.3 Digital filtering of the initial data

Richardson's table of initial data was of limited geographical extent. The coverage is indicated in Fig. 6.9 on p. 110, which is reproduced from *WPNP*. Using the observational data compiled by Hugo Hergesell and analysed by Vilhelm Bjerknes, grid-point data over the larger area shown in Fig. 6.10 (p. 111) were extracted. The method of analysing these data was described in §6.4. The re-analysed values were replaced by Richardson's original values wherever the latter were available. The initial fields are depicted in Figs. 9.1 and 9.2. The pressures at the five standard interfaces have been converted to heights at the five standard pressure levels 200, 400, 600, 800 and 1000 hPa. The sea-level pressure is shown in Fig. 9.1(f). The momentum values in the five layers have been converted to wind vectors and are shown in Fig. 9.2. The stratospheric temperature is shown in panel (f) of this figure. Pressure is high over the north of the area, with a south-easterly flow in the central region, in agreement with Bjerknes' analysis.

A cursory visual inspection of the charts suggests that the mass and wind fields are in approximate geostrophic balance. However, closer examination reveals marked departures from balance, especially at the uppermost levels. We recall that the re-analyses of mass and wind were completely independent of each other. This was done deliberately, to avoid subconscious imposition of geostrophic balance by the analysts. There is a peculiar feature on the 1000 hPa height chart (Fig. 9.1(e)) that is noteworthy. In §6.4 we found a discrepancy (at the point 48.6°N, 5.0°E) between Richardson's tabulated pressure and the re-analysed value, amounting to some 10 hPa. The anomalous 'bulls-eye' near Strasbourg confirms that Richardson's value at this point is in error. A similar anomaly appears in the sea-level pressure field but there is no corresponding distortion of the flow in the low-level winds (Fig. 9.2(e)).

The analysed fields were initialised by performing two integrations, one forwards and one backwards, each of duration $1\frac{1}{2}$ hours, and processing the values with a digital filter as described in §8.5. The time step for these integrations was

Initial fields, 0700 UTC, 20 May 1910.

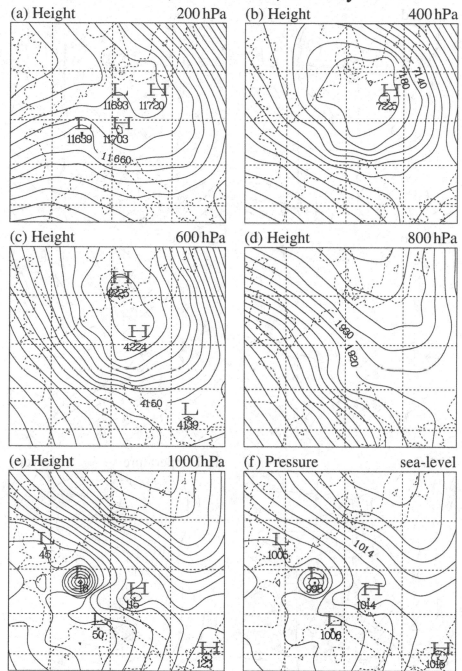

(a) Height   200 hPa

(b) Height   400 hPa

(c) Height   600 hPa

(d) Height   800 hPa

(e) Height   1000 hPa

(f) Pressure   sea-level

Figure 9.1 Initial fields of geopotential height at five standard levels and of sea-level pressure. Values in the central region agree with Richardson's tabulated values.

Initial fields, 0700 UTC, 20 May 1910.

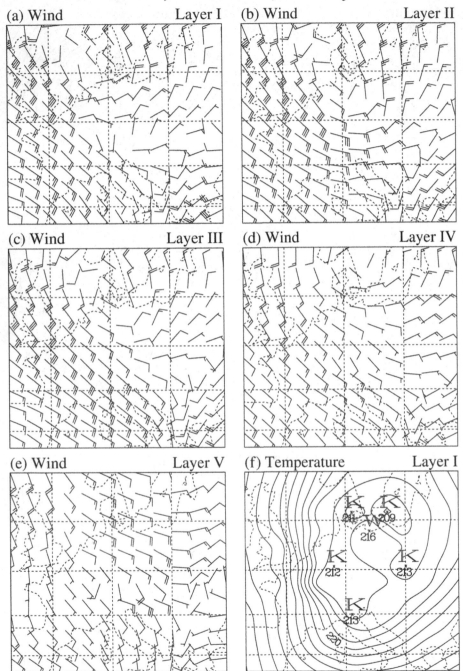

Figure 9.2 Initial wind fields in the five standard layers and stratospheric temperature. Values in the central region agree with Richardson's tabulated values.

Filtered fields, 0700 UTC, 20 May 1910.

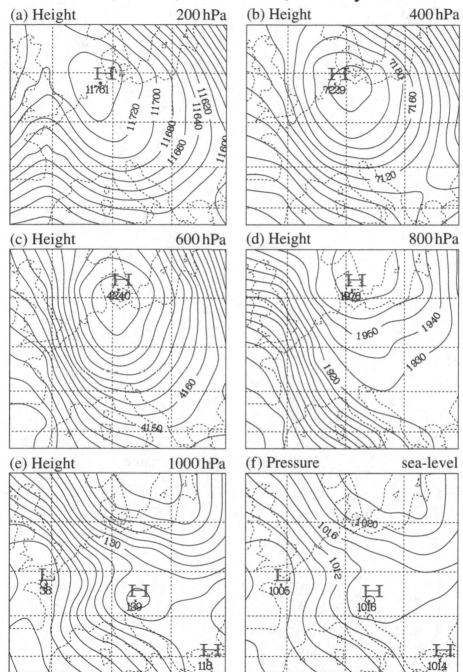

(a) Height   200 hPa

(b) Height   400 hPa

(c) Height   600 hPa

(d) Height   800 hPa

(e) Height   1000 hPa

(f) Pressure   sea-level

Figure 9.3 Filtered fields of geopotential height at five standard levels and of sea-level pressure. Details of the digital filtering procedure are given in the text.

Filtered fields, 0700 UTC, 20 May 1910.

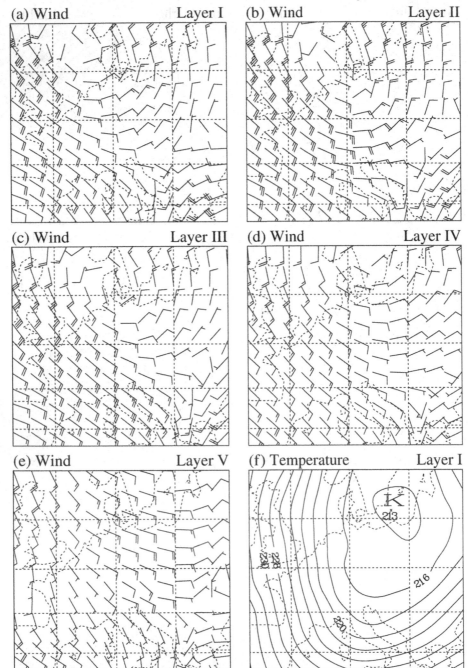

Figure 9.4 Filtered wind fields in the five standard layers and stratospheric temperature. Details of the digital filtering procedure are given in the text.

$\Delta t = 300$ s and the response function of the digital filter was shown in Fig. 8.5. Further details of the numerical integration scheme will be given below. The filtered fields are shown in Figs. 9.3 and 9.4. The overall pattern is maintained, with high pressure in the north and a south-easterly flow over the centre of the region. The fields are somewhat smoother than the original analysis. The anomalous bulls-eye over Strasbourg is completely absent in the filtered data (Fig. 9.3(e),(f)). This is reassuring: with no corresponding wind perturbation, this anomaly would necessarily be far from balance, and should be removed by filtering. However, we must remark that the anomaly was sufficiently displaced from the centre of the region that it was *not* a causative factor in Richardson's unrealistic value for the initial pressure tendency.

9.3.1 Initial tendencies after digital filtering

The predicted six-hour changes in pressure-thickness and momentum for each stratum are presented in Table 9.3. Results using Richardson's initial values are in columns 2, 3 and 4 and results for a forecast from initialised fields in columns 5, 6 and 7. The mean absolute values for each vertical column are given in the final row of the table. The characteristic sizes of the layer pressure changes for Richardson's data are about 30 times larger than those for the initialised data. For the momenta, Richardson's predicted changes are typically twice the magnitude of those for the filtered fields. We recall, from the analysis in Chapter 7, two causes of the unrealistic pressure tendencies: the divergence values were too large and there was a lack of compensation between convergence and divergence in the vertical. We re-examine this in detail now for the filtered fields.

The filtered pressure tendencies

Table 9.4, constructed using the filtered initial values, shows the total divergence and its components. The eastward and northward components are given in columns 2 and 3 and their sum in column 4. The numbers in the bottom row of the table are the mean absolute averages of the layer values and indicate the characteristic magnitudes of the terms. We see that there is some cancellation between the x- and y-components of divergence but it is very incomplete: the characteristic size of the horizontal divergence is $O(10^{-2}$ SI units), comparable to the size of the components themselves. This is of the same order of magnitude as for the uninitialised data (see Table 7.4 on page 127). It implies horizontal velocity divergence of magnitude $O(10^{-5}$ s$^{-1})$, which is quite large. The vertical component of divergence (Table 9.4, column 5) is comparable in magnitude to the horizontal components, and also to the uninitialised values. However, there is a highly significant difference for the filtered data: the horizontal and vertical components are approximately equal in magnitude

Smoothing the forecast

Table 9.3 *Predicted six-hour changes in pressure-thickness (ΔP) and in momentum (ΔU and ΔV) for each stratum using Richardson's initial values and values after initialisation with a digital filter.*

The mean absolute values for the column are shown in the final row. Pressure changes are in hPa and momentum changes in $\mathrm{kg\,m^{-1}s^{-1}}$.

	Before initialisation			After initialisation		
	ΔP	ΔU	ΔV	ΔP	ΔU	ΔV
I	48.3	−73 000	−33 700	−0.2	+50 500	+7200
II	28.7	−19 600	+23 800	−2.4	+6300	+12 800
III	26.2	−8900	+13 800	−0.4	+7900	+6500
IV	23.3	−15 300	−4300	−0.1	+6600	−500
V	18.6	−17 900	+6300	+2.1	−4700	+20 400
Mean abs. value	29.0	26 940	16 380	1.0	15 200	9480

Table 9.4 *Analysis of the components of the mass divergence for the filtered data.*

The final row gives the mean absolute values for each column and $[\rho w]$ denotes a vertical difference across the layer. All values are in SI units ($\mathrm{kg\,m^{-2}\,s^{-1}}$). This table should be compared to Table 7.4 on page 127.

	Eastward component	Northward component	Horizontal divergence	Vertical divergence	Total divergence
Layer	$\dfrac{\partial U}{\partial x}$	$\dfrac{\partial (V \cos \phi)}{\cos \phi \, \partial y}$	$\nabla \cdot \mathbf{U}$	$[\rho w]$	$\nabla \cdot \mathbf{U} + [\rho w]$
I	+0.0069	−0.0147	−0.0078	+0.0079	+0.0001
II	+0.0264	−0.0100	+0.0164	−0.0153	+0.0011
III	+0.0163	−0.0215	−0.0052	+0.0054	+0.0002
VI	+0.0079	−0.0022	+0.0056	−0.0056	+0.0000
V	−0.0121	+0.0035	−0.0086	+0.0075	−0.0010
Mean abs. value	0.0139	0.0104	0.0087	0.0083	0.0005

but opposite in sign, so there is substantial cancellation between them. As a result, the three-dimensional divergence for the filtered data (Table 9.4, column 6) is very much reduced in magnitude. It's characteristic size (the bottom right-hand entry in Table 9.4) is about 4 per cent of the corresponding value for the original data

Table 9.5 *Analysis of the pressure changes (hPa) for the filtered data.*

Changes $[p]$ across each layer and changes at the base of each layer are given. Values in column 4 are the products of those in column 3 by $-g\Delta t$, with $\Delta t = 6\,\text{h}$. Values should be compared to those in Table 7.5 on page 128.

Layer	Level	Total divergence $\nabla \cdot \mathbf{U} + [\rho w]$	Change in pressure thickness $\dfrac{\partial[p]}{\partial t}\Delta t$	Change in base pressure $\dfrac{\partial p}{\partial t}\Delta t$
I		+0.0001	−0.2	
	1			−0.2
II		+0.0011	**−2.4**	
	2			−2.6
III		+0.0002	−0.4	
	3			−3.0
IV		+0.0000	−0.1	
	4			−3.1
V		−0.0010	**+2.1**	
	S			−0.9

(Table 7.4). This accounts for the drastic reduction in the pressure-thickness changes after filtering that we saw in Table 9.3.

The pressure changes across each layer and the pressure change at the base of each layer are given in Table 9.5. There is a further critical distinction between the two initial data sets, made evident by a comparison between this table and Table 7.5 on page 128. Without initialisation, the presure changes are positive in all layers. Thus the vertical integral of divergence is large. We see in Table 9.5 that the largest pressure changes for the filtered data are in Layer II (−2.4 hPa/6 h) and Layer V (+2.1 hPa/6 h). They correspond to convergence near the ground and divergence in the upper troposphere. Since there is cancellation, the vertical integral of divergence is small. This illustrates the Dines compensation mechanism and, as a consequence of it, the surface pressure change is smaller than the changes across the layers.

There are no pressure observations available for the P-point of Richardson's forecast. However, observations for a number of stations in Bavaria were published by Schmauss (1911) in the *Deutsches Meteorologisches Jahrbuch für 1910.* The surface pressure at Bayreuth (49°57′ N, 11°34′ E, elevation 363.3 m) for a week in May 1910 is plotted in Fig. 9.5. It indicates that the pressure was quite steady at the initial time of the forecast, and only began to rise slowly the following day. Thus, the prediction from the filtered data of a slight fall in pressure is consistent with the available observations.

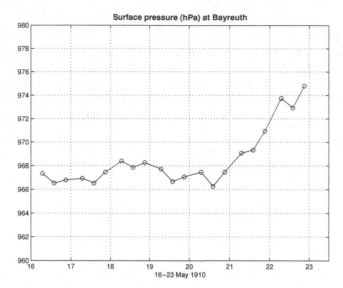

Figure 9.5 Surface pressure at Bayreuth (49°57′ N, 11°34′ E, Elevation 363.3 m) for a week in May, 1910. Pressures in mm Hg converted to hPa by multiplication by $\frac{4}{3}$. Observations are at 0700, 1400 and 2100 UTC each day. (Data from Schmauss, 1911)

The filtered momentum tendencies

For balanced atmospheric flow, the tendencies arise as small residual differences between large quantities. The principal terms of the eastward momentum equation for the filtered date are tabulated in Table 9.6. The pressure gradient and Coriolis terms dominate, and there is some tendency for them to cancel; however, this cancellation is far from complete. The numbers in the bottom row of the table are the mean absolute averages of the layer values. They show that the tendency of U is comparable in size to the pressure and Coriolis terms. Thus, despite the filtering, the fields are not particularly close to geostrophic balance. Moreover, the U-tendency in the stratosphere remains surprisingly high. However, the overall size of the tendencies is about half that for uninitialised data (Table 7.7, page 130). The results for the meridional momentum component (Table 9.7) confirm these findings.

In summary, for the filtered data the horizontal compensation remains weak, but the vertical compensation is strong and the Dines compensation mechanism is evident. There remains marked imbalance between the mass and wind fields, so that the momentum tendencies are reduced by only a factor of two from Richardson's values. Finally, the stratospheric temperature tendency for the filtered data is $2.2 \times 10^{-4}\,\mathrm{K\,s^{-1}}$, or 4.8 K in six hours. This is large, but much smaller than the value of 19.6 K in six hours for the unfiltered data.

Table 9.6 *Terms in eastward momentum equation and tendency of eastward momentum (kg m^{-1} s^{-2}).*

All values are for the filtered data. This table should be compared to Table 7.7 on page 130.

Layer	Pressure gradient term $-\dfrac{\partial P}{\partial x}$	Coriolis term $+fV$	Remaining terms —	Tendency of momentum $\dfrac{\partial U}{\partial t}$	Ageostrophic wind $f(V - V_{\text{geo}})$
I	+3.05	−0.78	+0.07	+2.34	+2.26
II	+0.64	−0.61	+0.26	+0.29	+0.03
III	+0.09	+0.16	+0.11	+0.37	+0.25
IV	−0.16	+0.45	+0.01	+0.31	+0.29
V	−0.77	+0.58	−0.03	−0.22	−0.19
Mean abs. value	0.94	0.52	0.10	0.70	0.60

Table 9.7 *Terms in northward momentum equation and tendency of northward momentum (kg m^{-1} s^{-2}).*

All values are for the filtered data. This table should be compared to Table 7.8 on page 130.

Layer	Pressure gradient term $-\dfrac{\partial P}{\partial y}$	Coriolis term $-fU$	Remaining terms —	Tendency of momentum $\dfrac{\partial V}{\partial t}$	Ageostrophic wind $f(U - U_{\text{geo}})$
I	−0.63	+1.07	−0.11	+0.33	+0.44
II	−0.74	+1.73	−0.40	+0.59	+0.99
III	−0.91	+1.38	−0.17	+0.30	+0.47
IV	−0.88	+0.91	−0.06	−0.02	+0.03
V	−0.46	+1.42	−0.01	+0.94	+0.12
Mean abs. value	0.72	1.30	0.15	0.44	0.41

9.3.2 Short-range forecasts

To obtain a more synoptic view of the pressure tendencies, two one-hour forecasts were made, one from the original and one from the filtered data. Each forecast comprised a single forward step of size $\Delta t = 3600$ s. That is, the computed tendencies were multiplied by Δt and added to the initial fields. The two one-hour 'forecasts' of sea-level pressure are shown in Fig. 9.6. The forecast from the original data

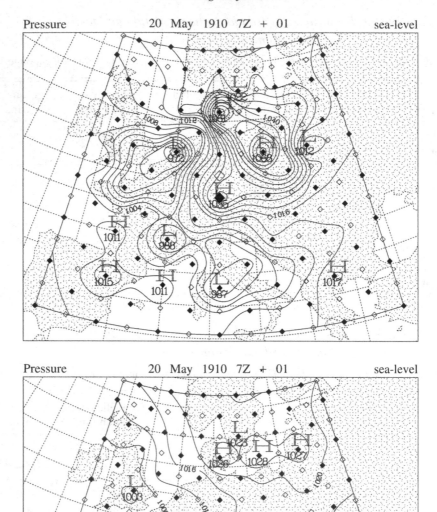

Figure 9.6 Two short-range forecasts of sea-level pressure, from (top) uninitialised data and (bottom) filtered data. The contour interval is 4 hPa for both charts. A single forward time step of size $\Delta t = 3600$ s was made in each case.

(upper panel) is outlandish. At the centre point, the pressure has risen by 25 hPa. The extreme pressure fall and rise are -36 hPa and $+51$ hPa respectively. For the filtered data, the pressure at the centre point is almost steady, which is reasonable. There is some residual noise evident near the edges of the domain, with a maximum pressure change of 6 hPa. This unrealistic value is a result of the ineffectiveness of the digital filtering near the boundary of the region; for operational application of DFI there are means of rectifying this defect. The rms (root mean square) change in pressure averaged over the forecast area is more than 10 hPa for the uninitialised data; for the filtered data, it is about 1 hPa.

9.4 Extension of the forecast

9.4.1 Horizontal diffusion

Two 24-hour forecasts were made, one starting from Richardson's data and one from the filtered data. For both integrations the time step was set to $\Delta t = 300$ s. This is well within the limitations imposed by the Courant–Friedrichs–Lewy criterion. Nevertheless, preliminary integrations suffered from catastrophic numerical instabilities and failed before the target time. In order to maintain numerical stability, various smoothing methods were investigated. A Robert–Asselin time filter (Asselin, 1972) was tested. While it had a minor beneficial effect in reducing $2\Delta t$ noise, it did not stabilise the integrations. Divergence damping (Sadourny, 1975) was tested, but also failed to stabilise the runs. Finally, spatial diffusion of the fields was applied during the integrations. This was done using a sixth-order low-pass implicit tangent filter devised by Raymond (1988). Extensive numerical experimentation confirmed that spatial smoothing was essential and that a coefficient of $\epsilon_S = 0.1$ was about optimal. With the Raymond filter active, the divergence damping and the Robert–Asselin time filter were disabled, so that the only explicit damping was horizontal diffusion.

The idea of using smoothing during the integration was put forward by Richardson as Method E in §9.2 above. He was concerned with removal of small-scale noise but was not aware of the hazard of computational instability. An essential feature of finite difference schemes is their inability to properly represent physical phenomena with spatial scales smaller than the grid size. When the equations are averaged over a grid cell, additional eddy-diffusion terms arise. These have not been accounted for in the model used in this study. Moreover, the cascade of energy to small scales, where dissipation processes act, cannot be properly represented in the coarse grid model. Thus, there is a tendency for an artificial build-up of energy at the scale of the grid, and it is necessary to incorporate a mechanism by which this spurious accumulation of energy can be prevented. The management of small-scale noise has been a continuing problem throughout the history of computer forecasting. The

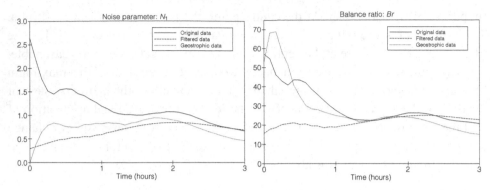

Figure 9.7 Evolution of two gravity wave noise indicators. Left panel: mean absolute surface pressure tendency N_1 (units Ber or hPa/h). Right panel: balance ratio Br (per cent). Solid lines: forecast from uninitialised data. Dashed lines: forecast from filtered data. Dotted lines: forecast from geostrophic data. The first three forecast hours are shown.

requirement for numerical diffusion, and the construction of appropriate diffusion schemes has been one of the major challenges facing modellers.

9.4.2 Noise indicators

To compare the two integrations, a number of indicators of high-frequency oscillations were computed. The primary measure of gravity-wave noise is the mean absolute surface pressure tendency, designated N_1 and defined by

$$N_1 = \frac{1}{IJ} \sum_{i,j} \left| \frac{\partial p_S}{\partial t} \right|_{ij} = \frac{g}{IJ} \sum_{i,j} \left| \sum_{k} (\nabla \cdot \mathbf{V})_{ijk} \right|. \qquad (9.5)$$

The evolution of the noise indicator N_1 for the first three hours of the two forecasts is shown in Fig. 9.7, left panel (a third integration, using geostrophic initial winds, will be considered below). For the uninitialised data, N_1 starts at the unrealistically large value of 2.5 Ber (or 2.5 hPa per hour) and takes about three hours to fall to a steady value around 0.75 Ber. For the filtered data, the initial value of N_1 is much smaller, and it remains fairly steady through the forecast.

Another measure, more sensitive to *internal* gravity wave oscillations, is the mean absolute divergence N_2, defined by

$$N_2 = \frac{g}{IJ} \sum_{i,j} \sum_{k} |(\nabla \cdot \mathbf{V})_{ijk}|. \qquad (9.6)$$

The Dines compensation effect may be evaluated by defining the *balance ratio*

$$Br = 100 \left(\frac{N_1}{N_2} \right). \qquad (9.7)$$

The factor of 100 conveniently allows Br to be expressed as a percentage of its maximum possible value. The evolution of this quantity is shown in Fig. 9.7 (right panel). For the original data, its initial value is about 57 per cent, and the ratio falls to about 25 per cent after three hours. For the filtered data the evolution of the balance ratio is much more steady, remaining in the interval 15–25 per cent through the run.

Geostrophic initial winds

Nondivergent initial winds can be defined in a number of ways. For example, the technique of partitioning the wind into rotational and solenoidal components described in Lynch (1989) could have been used to compute a stream function for the non-divergent component of the analysed wind. However, since the analysed mass and wind fields are drastically out of balance, this will still be the case even if the divergence is removed. It is more interesting to use a nondivergent wind that is compatible with the pressure field. This can easily be done by deriving the geostrophic wind from the analysed pressure. If a constant mean value f_0 of the Coriolis parameter is used, the resulting wind will be nondivergent. Thus, we define

$$u_{geo} = -\frac{1}{f_0 a}\frac{\partial \bar{p}}{\partial \phi}, \qquad v_{geo} = +\frac{1}{f_0 a \cos\phi}\frac{\partial \bar{p}}{\partial \lambda},$$

where $\bar{p}$ represents the pressure at the centre of the stratum. This amounts to treating $\bar{p}/f_0$ as the stream function and assuming that the velocity potential vanishes, $\chi \equiv 0$. Obviously, since

$$\nabla \cdot \mathbf{V}_{geo} = \frac{1}{a \cos\phi}\left[\frac{\partial u_{geo}}{\partial \lambda} + \frac{\partial \cos\phi\, v_{geo}}{\partial \phi}\right] \equiv 0,$$

the noise parameters N_1 and N_2 are zero at the initial time when such geostrophic initial winds are used. The evolution of N_1 and Br through the three-hour forecast are shown in Fig. 9.7. While N_1 remains reasonably small through the forecast, the balance ratio Br actually reaches a value of almost 70 per cent, which is higher than the values for the uninitialised forecast. The enforcement of nondivergence at the initial time results in a reasonable evolution of the surface pressure field. However, the Dines compensation mechanism is not well represented in the initial evolution from geostrophic initial fields.

9.4.3 Effect of filtering on the forecast

The 24-hour forecasts from the original and filtered data were compared to assess the impact of the initialisation. Ideally, the filtering should remove high frequency noise but should not otherwise have a significant impact on the forecast. In Table 9.8,

Table 9.8 *Maximum and root mean square differences between the filtered and unfiltered fields of height and winds in each layer and of surface pressure at the initial time, after six hours and after a one-day forecast.*

Height differences are in metres, winds in ms^{-1} and pressures in hPa.

		At initial time		After 6 hours		After 24 hours	
		Max.	RMS	Max.	RMS	Max.	RMS
I	Δz	71.499	17.228	3.150	1.406	0.252	0.114
	Δu	5.384	1.226	0.162	0.037	0.004	0.001
	Δv	7.155	1.821	0.250	0.107	0.005	0.002
II	Δz	34.790	10.055	4.523	1.661	0.147	0.065
	Δu	5.023	1.046	0.552	0.200	0.032	0.012
	Δv	5.535	1.184	0.466	0.153	0.036	0.013
III	Δz	26.943	8.900	2.112	0.820	0.050	0.019
	Δu	5.059	0.838	0.191	0.065	0.005	0.002
	Δv	3.072	0.834	0.264	0.084	0.012	0.005
VI	Δz	28.996	8.606	1.670	0.461	0.053	0.018
	Δu	4.023	0.837	0.272	0.096	0.013	0.006
	Δv	3.286	0.701	0.136	0.039	0.017	0.007
V	Δz	87.952	15.341	4.312	2.008	0.413	0.168
	Δu	6.349	0.983	0.833	0.150	0.037	0.009
	Δv	5.614	0.826	0.956	0.174	0.023	0.006
	Δp_s	10.428	1.819	0.511	0.238	0.049	0.020

the maximum and root mean square differences in height Δz and winds Δu and Δv for each layer, and for the surface pressure Δp_s are shown for the initial data, after 6 hours and after 24 hours. We see, for example, that filtering has resulted in a maximum change of more than 10 hPa. However, the differences after 24 hours are so small as to be completely negligible. Thus, the digital filtering of the initial data does not have any deleterious effect on the forecast.

The initial sea-level pressure field and the 24-hour forecast (filtered case) are shown in Fig. 9.8. During the course of the integration, the pressure has risen over the central region by about 5 hPa. This is completely consistent with the observed rise in pressure at Bayreuth, which was shown in Fig. 9.5 (p. 172). It is encouraging that the one-day forecast is so reasonable. It must be remarked, however, that the geographical extent of the forecast domain is so limited that boundary effects must soon begin to dominate the evolution. Since the boundary values are held constant,

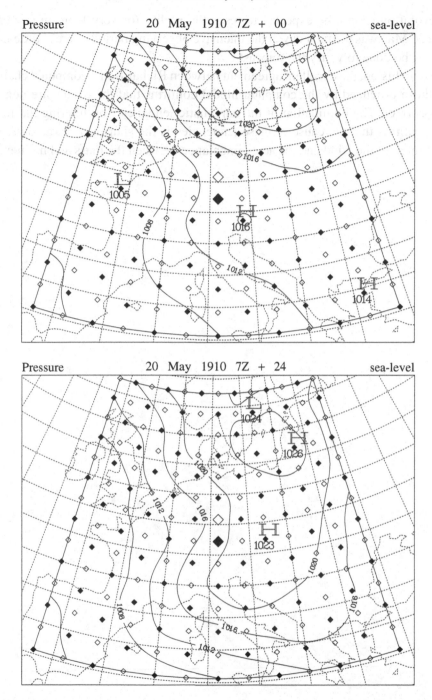

Figure 9.8 Analysis (upper panel) and 24-hour forecast (lower panel) sea-level pressure fields based on filtered initial data. The contour interval is 4 hPa for both charts.

the evolution cannot be expected to resemble reality for very long. In any case, since 20 May was the final day of the intensive observing period, there is little more that can be done to verify the forecast.

In conclusion, the excellent agreement between Richardson's computed changes and those produced by the computer model confirms that the results are numerically correct. The elimination of the unrealistic tendencies by filtering the initial data confirms that the cause of these tendencies is the spuriously large divergent component of the wind and the acute imbalance between the mass and momentum fields.

10

The ENIAC integrations

> The role of the enormous weather factory envisaged by Richardson (1922)
> with its thousands of computers will ... be taken over by a completely
> automatic electronic computing machine. *(Charney, 1949)*

It is without question that Richardson's attempt to predict the weather by numerical means, while visionary and courageous, was premature. In his review of *WPNP*, Exner (1923) expressed the view that Richardson's method was unlikely to lead to progress in weather forecasting and even recommended that he should write a book on theoretical meteorology that was free from the aim of a direct application to prediction. This negative response was echoed by several other reviewers. Later, the renowned meteorologist Bernhard Haurwitz wrote in his textbook *Dynamic Meteorology* that efforts to compute weather changes by direct application of the equations were not promising, adding that

> ... a computation of the future weather by dynamical methods will be possible only when it
> is known more definitely which factors have to be taken into account under given conditions
> and which may be neglected. *(Haurwitz, 1941, p. 180)*

Haurwitz understood that a simplification of the mathematical formulation of the forecasting problem was required but he was not in a position to propose any specific solution. In fact, there were several obstacles preventing the fulfilment of Richardson's dream, and progress was required on four separate fronts before it could be realised.

Firstly, in order to develop a simplified system suitable for numerical prediction, a better understanding of atmospheric dynamics, and especially of the wave motions in the upper atmosphere, was required. Major insights were provided by the work of Jacob Bjerknes and Reginald Sutcliffe, amongst many others, and especially by Rossby's mechanistic description – powerful in its simplicity – of atmospheric waves. This was followed by Charney's (1947) explanation of cyclonic development in terms of baroclinic instability and Kuo's (1949) work on barotropic instability. Secondly, the development of the radiosonde made observations of the free atmosphere possible in real time. A network of surface and upper air stations, established to support military operations during World War II, was developed and strengthened to serve the needs of civil aviation. Thus, it became possible to construct a comprehensive synoptic description of the state of the atmosphere. Thirdly, an understanding of the stability properties of finite difference schemes flowed from the work of Courant *et al.* (1928) in Göttingen. This provided the key to ensuring that the numerical solution was a reasonable representation of reality. And fourthly, the development of automatic electronic computing machinery provided a practical means of carrying out the monumental computational task of calculating changes in the weather.

10.1 The 'Meteorology Project'

John von Neumann

John von Neumann was one of the leading mathematicians of the twentieth century. He made important contributions in several areas: mathematical logic, functional analysis, abstract algebra, quantum physics, game theory and the development and application of computers. A brief sketch of his life may be found in Goldstine (1972) and several biographies have been written (e.g. Heims, 1980; Macrae, 1999). Von Neumann was born in Hungary in 1903. He showed outstanding intellectual and linguistic ability at an early age. After studying in Budapest, Zurich and Berlin he spent a period in the 1920s working in Göttingen with David Hilbert on the logical foundations of mathematics. In 1930 he was invited to Princeton University, and he remained at the Institute for Advanced Studies for 25 years. He died in 1957, aged only 54. In the mid 1930s von Neumann became interested in turbulent fluid flows. The nonlinear partial differential equations that describe such flows defy analytical assault and even qualitative insight comes hard. Von Neumann was a key figure in the Manhattan Project, which led to the development of the atomic bomb. This project involved the solution of hydrodynamic problems vastly more complex than had ever been tackled before. Von Neumann was acutely aware of the difficulties and limitations of the available solution methods:

'Our present analytical methods seem unsuitable for the important problems arising in connection with the solution of non-linear partial differential equations and, in fact, with all types of non-linear problems of pure mathematics. The truth of this statement is particularly striking in the field of fluid dynamics. Only the most elementary problems have been solved analytically in this field'.   *(in Goldstine, 1972, pp. 179–80)*

Von Neumann saw that progress in hydrodynamics would be greatly accelerated if a means of solving complex equations numerically were available. It was clear that very fast automatic computing machinery was required. He masterminded the design and construction of an electronic computer at the Institute for Advanced Studies. This machine was built between 1946 and 1952 and its design had a profound impact upon the subsequent development of the computer industry. This *Electronic Computer Project* was 'undoubtedly the most influential single undertaking in the history of the computer during this period' (Goldstine, 1972, p. 255). The Project comprised four groups: (1) Engineering, (2) Logical Design and Programming, (3) Mathematical, and (4) Meteorological. The fourth group was directed by Jule Charney from 1948 to 1956.

Von Neumann was legendary for his astounding memory, his capacity for mental calculation at lightning speed and his highly developed sense of humour. Goldstine speaks of the guidance and help that he so freely gave to his friends and aquaintances, both contemporary and younger than himself. He was described by many who knew him as a man of great personal charm, and numerous anecdotes attest to his unique genius. Von Neumann and Richardson could hardly have been more different in personality or outlook. Von Neumann was socially sophisticated, extrovert and at complete ease in company, with a vast repertoire of amusing stories, limericks and jokes. Richardson was reserved and withdrawn, most comfortable when alone and even somewhat stand-offish. He once proposed listing 'solitude' as a hobby in his entry for *Who's Who* (Ashford, 1985, p. 175). Politically, the contrast was even starker: as we have seen, Richardson was a committed and immovable pacifist; von Neumann advocated preventative war, favouring a pre-emptive nuclear strike against the Soviet Union. It is fortunate that his views on this issue did not prevail.

The 'Conference on Meteorology'

Von Neumann recognised weather forecasting, a problem of both great practical significance and intrinsic scientific interest, as a problem par excellence for an automatic computer. His work at Los Alamos on hydrodynamic problems had given him a profound understanding of the difficulties in this area.

Moreover, he knew of the pioneering work ... [of] Lewis F Richardson.... [which] failed largely because the Courant condition had not yet been discovered, and because high speed computers did not then exist. But von Neumann knew of both. *(Goldstine, 1972, p. 300)*

Von Neumann had been in Göttingen in the 1920s when Courant, Friedrichs and Lewy were working on the numerical solution of partial differential equations and he fully apreciated the practical implications of their findings. However, Goldstine's suggestion that the Courant–Friedrichs–Lewy (CFL) criterion was responsible for

Richardson's failure is wide of the mark. This erroneous explanation has also been widely promulgated by others.

Von Neumann made estimates of the computational power required to integrate the equations of motion of the atmosphere and concluded tentatively that it would be feasible on the Institute of Advanced Studies' (IAS) computer (popularly called the *Johnniac*). A formal proposal was made to the US Navy to solicit financial backing for the establishment of a Meteorology Project. According to Platzman (1979) this proposal was 'perhaps the most visionary prospectus for numerical weather prediction since the publication of Richardson's book a quarter-century earlier'. Its purpose is stated at the outset (the full text is reproduced in Thompson, 1990):

The objective of the project is an investigation of the theory of dynamic meteorology in order to make it accessible to high speed, electronic, digital, automatic computing, of a type which is beginning to become available, and which is likely to be increasingly available in the future.

Several problems in dynamic meteorology were listed in the proposal. It was clear that, even if the computer were available, it could not be used immediately. Some theoretical difficulties remained, which could only be overcome by a concerted research effort: this should be done by 'a group of five or six first-class younger meteorologists'. The possibilities opened up by the proposed project were then considered:

Entirely new methods of weather prediction by calculation will have been made practical. It is not difficult to estimate that with the speeds indicated ... above, a completely calcu- lated prediction for the entire northern hemisphere should take about two hours per day of prediction.

Other expected benefits were listed, including advances towards 'influencing the weather by rational, human intervention ... since the effects of any hypothetical intervention will have become calculable'. The proposal was successful in attracting financial support, and the Meteorological Research Project began in July 1946.

A meeting – the Conference on Meteorology – was arranged at the Institute the following month to enlist the support of the meteorological community and many of the leaders of the field attended. Von Neumann had discussed the prospects for numerical weather forecasting with Carl Gustaf Rossby. Indeed, it remains unclear just which of them first thought of the whole idea. Von Neumann had tried to attract Rossby to the Institute on a permanent basis but succeeded only in bringing him for short visits. Having completed a brilliant PhD thesis on the baroclinic instability of the westerlies, Jule Charney stopped off on his way to Norway, to visit Rossby in Chicago. They got on so well that the three-week stay was extended to almost a year. Charney described that spell as 'the main formative experience of my whole professional life' (Platzman, 1990, referenced below as *Recollections*). As an inducement to Charney to stay in Chicago, Rossby arranged for him to be invited to the Princeton meeting. Charney was at that time already somewhat familiar with Richardson's book. Richardson's forecast was much discussed at the meeting. It was clear that the CFL stability criterion prohibited the use of a long time step such as had been used in *WPNP*. The initial plan was to integrate the primitive equations; but the existence of high-speed gravity wave solutions required the use of such a short time step that the volume of computation might exceed the capabilities of the IAS machine. And there was a more fundamental difficulty: the impossibility of accurately calculating the divergence from the observations. Charney believed that Richardson's fundamental error was that

his initial tendency field was completely wrong because he was not able to evaluate the divergence. He couldn't have used anything better than the geostrophic wind... [which] would have given the false divergence. I thought that maybe the primitive equations were just not appropriate.   *(Recollections, p. 39)*

(Recall how it was shown in Chapter 3 that the divergence of the geostrophic wind does not faithfully reflect that of the flow). Thus, two obstacles loomed before the participants at the meeting: how to avoid the requirement for a prohibitively short time step, and how to avoid using the computed divergence to calculate the pressure tendency. The answers were not apparent; it remained for Charney to find a way forward.

Jule Charney

Jule Charney (Fig. 10.1) was born on New Year's Day 1917, in San Francisco, to Russian Jewish parents. He studied mathematics at UCLA, receiving a bachelor's degree in 1938. Shortly afterwards, Jacob Bjerknes organised a programme in meteorology in Los Angeles, and Charney became a teaching assistant. The great unsolved problem at that time was the genesis and development of extratropical depressions. The Norwegian scientists Bjerknes, Holmboe and Solberg had studied this problem but with inconclusive results. Charney's starting point was a wave perturbation on a zonally symmetric basic state with westerly flow and north–south temperature gradient. He reduced the problem to a second order ordinary differential equation, the confluent hypergeometric equation. He found solutions that, for sufficiently strong temperature gradients, grew with time, and thereby explained cyclonic development in terms of baroclinic instability of the basic state. His report (Charney, 1947) took up an entire issue of the *Journal of Meteorology* and was instantly recognised as of fundamental importance.

Charney's many contributions to atmospheric dynamics, oceanography and international meteorology are described in Lindzen *et al.* (1990). We focus here on

Figure 10.1 Jule G. Charney (1917–1981). From the cover of *Eos*, Vol. 57, August, 1976. (© Nora Rosenbaum)

his role in the emergence of numerical weather prediction. His key paper 'On a physical basis for numerical prediction of large-scale motions in the atmosphere' (Charney, 1949) addresses some crucially important issues. Charney considered the means of dealing with high frequency noise, proposing a hierarchy of filtered models, which we will discuss shortly. Using the concept of group velocity, he investigated the rate of travel of meteorological signals and concluded that, with the data coverage then available, numerical forecasts for one, or perhaps two, days were possible for the eastern United States and Europe. Finally, Charney presented in this paper the results of a manual computation of the tendency of the 500 hPa height which he and Arnt Eliassen had made using the barotropic vorticity equation. To solve the associated Poisson equation, they used a relaxation method originally devised by Richardson. The results are shown in Fig. 10.2. As the verifying analysis was available only between 140°W and 20°E, we show only a detail of the figure in

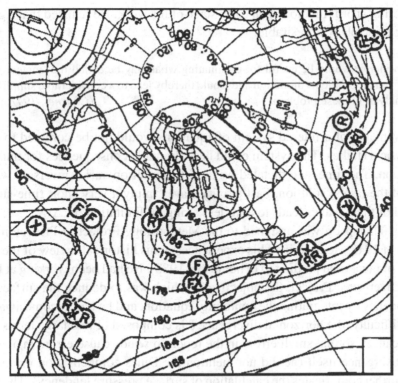

Figure 10.2 Manual integration of the barotropic vorticity equation. R and R'
indicate points of maximum *observed* and *computed* height rise. F and F' indicate
points of maximum *observed* and *computed* height fall. Xs mark the extrema of
vorticity advection. (Detail of figure from Charney (1949), © Amer. Met. Soc.)

Charney (1949). The letters R and F indicate points of maximum *observed* height
rise and fall, and R' and F' the corresponding points for the *computed* tendencies.
The agreement is impressive. The Xs mark the extrema of vorticity advection. The
centres of rise and fall are located approximately at these points.

10.2 The filtered equations

In his baroclinic instability study, Charney had derived a mathematically tractable
equation for the unstable waves 'by eliminating from consideration at the outset
the meteorologically unimportant acoustic and shearing-gravitational oscillations'
(Charney, 1947). He considered linear perturbations of infinite lateral extent and
removed the high-speed waves by assuming certain quantities (related to the internal
and external Froude numbers $c^2/\Re T$ and c^2/gH) were small. Thus, he had the idea

of a filtering approximation at an early stage of his work. He realised that a general filtering principle was desirable:

Such a principle would be useful for eliminating what may be called the 'meteorological noises' from the problems of motion and would thereby lead to a considerable simplification of the analysis of these problems.   *(Charney, 1947, p. 234).*

The advantages of a filtered system of equations would not be confined to its use in analytical studies. The system could have dramatic consequences for numerical integration. The time-step dictated by the CFL criterion varies inversely with the speed of the fastest solution. Fast gravity waves imply a very short time-step; the removal of these waves leads to a far less stringent limitation on Δt.

Philip Thompson was assigned to the Meteorology Project in Princeton in the autumn of 1946. He described his experiences in his historical review of the development of numerical prediciton (Thompson, 1983). He had been working at UCLA on the Divergence Project, the objective of which was to deduce the surface pressure tendency by integrating the continuity equation, and he had become aware of a major difficulty: the horizontal divergence δ is composed of two large terms which almost cancel, so that small errors in the reported winds cause errors in δ as large as the divergence itself (he did not mention the Dines compensation mechanism, which further complicates the calculation of surface pressure tendency). Thompson abandoned the original objectives of the Divergence Project and sought an alternative approach, in the course of which he derived a diagnostic equation for the vertical velocity – unaware that the same equation had been derived by Richardson some 30 years earlier. Thompson had met Charney when they were in Los Angeles. He wrote to him in Chicago in early 1947 asking him about some problems associated with gravity waves. In his response (dated 12 February 1947), Charney outlined his ideas about filtering the noise. First he explained that, since the primary excitation mechanisms in the atmosphere are of long period, the resulting disturbances are also of low frequency. To illustrate the situation, he employed the musical analogy quoted above (p. 138). He then sketched the derivation of the dispersion relation for the wave phase-speeds:

$$\bar{u} - c - \frac{\beta L^2}{4\pi^2} = \frac{L^2 f^2}{4\pi^2} \frac{c}{gH - (\bar{u} - c)^2}. \tag{10.1}$$

This cubic equation for c had appeared in his baroclinic instability study and also, much earlier, in Rossby (1939). The three roots are given approximately by

$$c_1 \approx \bar{u} - \frac{\beta L^2}{4\pi^2}, \qquad c_{2,3} \approx \bar{u} \pm \sqrt{gH}, \tag{10.2}$$

the slow rotational wave and the two gravity waves travelling in opposite directions. 'Since most of the energy of the initial disturbance goes into long period components, very little . . . will appear in the gravitational wave form'.

He then considered the question of how to filter out the noise. He drew an analogy between a forecasting model and a radio receiver, and argued that the noise could be either eliminated from the input signal or removed by a filtering system in the receiver. He described a method of filtering the equations in a particular case, but concluded 'I still don't know what types of approximation have to be made in more general situations.' It did not take him long to find out. In a second letter, dated 4 November the same year, he wrote:

The solution is so absurdly simple that I hesitate to mention it. It is expressed in the following principle. Assuming conservation of entropy and absence of friction in the free atmosphere, the motion of *large-scale* systems is governed by the laws of conservation of potential temperature and potential vorticity *and* by the condition that the field of motion is in hydrostatic and *geostrophic* balance. This is the required filter!

Charney's two letters are reproduced in Thompson (1990). A full account of the filtering method was published in the paper 'On the scale of atmospheric motions' (Charney, 1948); this paper was to have a profound impact on the subsequent development of dynamic meteorology. Charney analysed the primitive equations using the technique of scale analysis, which we have described above (§7.2). He was able to simplify the system in such a way that the gravity wave solutions were completely eliminated. The resulting equations are known as the *quasi-geostrophic* system. The system boils down to a single prognostic equation for the quasi-geostrophic potential vorticity,

$$\left(\frac{\partial}{\partial t} + \mathbf{V} \cdot \nabla\right)\left[f + \zeta + \frac{f_0}{\rho_0}\frac{\partial}{\partial z}\left(\frac{\rho_0}{N^2}\frac{\partial p/\rho_0}{\partial z}\right)\right] = 0, \qquad (10.3)$$

where N^2 is the Brunt–Väisälä frequency and p is the deviation from the reference pressure $p_0(z)$. The wind is assumed to be geostrophic and the vorticity is related to the pressure by $\zeta = (1/\rho_0 f)\nabla^2 p$. All that is required by way of initial data to solve this equation is a knowledge of the three-dimensional pressure field (and appropriate boundary conditions).

A filtered system quite similar in character to the quasi-geostrophic system, and subsequently christened the semi-geostrophic equations, was derived independently by Arnt Eliassen, by means of a substitution of the geostrophic approximation into the acceleration terms (this is the geostrophic momentum approximation). In the same volume of *Geofysiske Publikasjoner*, Eliassen (1949) presented the full system of equations of motion in isobaric co-ordinates. The idea of using pressure as the vertical co-ordinate has been of central importance in numerical modelling of the atmosphere.

In the special case of horizontal flow with constant static stability, the vertical variation can be separated out and the quasi-geostrophic potential vorticity equation reduces to a form equivalent to the nondivergent barotropic vorticity equation

$$\frac{d(f + \zeta)}{dt} = 0. \tag{10.4}$$

In fact, Charney (1949) showed that, under less restrictive assumptions, the three-dimensional forecast problem may be reduced to the solution of a two-dimensional equation for an 'equivalent barotropic' atmosphere. Charney began a manual integration of this equation while he was still in Norway; the results of this were shown in Fig. 10.2 above. The barotropic equation had, of course, been used by Rossby (1939) in his analytical study of atmospheric waves, but nobody seriously believed that it was capable of producing a quantitatively accurate prediction of atmospheric flow. Charney now saw it as the first member of a hierarchy of models of increasing complexity and verisimilitude (Charney, 1949). Its position as a direct specialisation of the more general quasi-geostrophic equation made it more credible for use as a first test-case of numerical weather prediction.

Charney became the leader of the Meteorology Group in mid-1948 and remained until the termination of the project eight years later. Arnt Eliassen arrived slightly later than Charney. He returned to Oslo after a year and was replaced by Ragnar Fjørtoft. By the time Charney arrived in Princeton, he had the quasi-geostrophic equations 'in his pocket'. He also saw how to integrate them numerically: it was a matter of advecting the potential vorticity and then solving a Poisson equation for the stream-function. He felt that one should begin with the simplest model, the barotropic equation, and gradually introduce physical and mathematical factors one at a time. The intention was to progress rapidly to a baroclinic model, since the prediction of cyclogenesis was considered to be the central problem. In fact, the practical usefulness of the barotropic equation had been greatly underestimated: in *Recollections* (p. 49) he says 'I think we were all rather surprised that the predictions were as good as they were.' We will now describe the trail-blazing work that culminated in the successful numerical integration of that simple equation.

10.3 The first computer forecast

In 1948, the Meteorology Group adopted the general plan of attacking the problem of numerical weather prediction by investigating a hierarchy of models of increasing complexity, starting with the simplest, the non-divergent barotropic vorticity equation. By early 1950 they had completed the necessary mathematical analysis and had designed a numerical algorithm for solving this equation. The scientific record of this work is the much-cited paper in *Tellus*, by Charney, Fjørtoft and von Neumann

(1950). The authors outlined their reasons for starting with the barotropic equation: the large-scale motions of the atmosphere are predominantly barotropic; the simple model could serve as a valuable pilot-study for more complex integrations; and, if the results proved sufficiently accurate, barotropic forecasts could be utilised in an operational context. In fact, nobody anticipated the enormous practical value of this simple model and the leading role it was to play in operational prediction for many years to come (Platzman, 1979). As the IAS machine was still two years from completion, arrangements were made to run the integration on the only computer then available. The Electronic Numerical Integrator and Computer (ENIAC), which had been completed in 1945, was the first multi-purpose electronic digital computer ever built. It was installed at the Ballistic Research Laboratories at Aberdeen, Maryland. It was a gigantic contraption with 18 000 thermionic valves, massive banks of switches and large plugboards with tangled skeins of connecting wires, filling a large room and consuming some 140 kW of power. Program commands were specified by setting the positions of a multitude of ten-pole rotary switches on large arrays called function tables, and input and output was by means of punch-cards. The time between machine failures was typically a few hours, making the use of the computer a wearisome task for those operating it.

The numerical algorithm

The method chosen by Charney *et al.* to solve the barotropic vorticity equation

$$\frac{\partial \zeta}{\partial t} + \mathbf{V} \cdot \nabla(\zeta + f) = 0 \tag{10.5}$$

was based on using geopotential height as the prognostic variable. If the wind is taken to be both geostrophic and nondivergent, we have

$$\mathbf{V} = (g/f)\mathbf{k} \times \nabla z; \qquad \mathbf{V} = \mathbf{k} \times \nabla \psi.$$

The vorticity is given by $\zeta = \nabla^2 \psi$. These relationships lead to the linear balance equation

$$\zeta = g\nabla \cdot (1/f)\nabla z = (g/f)\nabla^2 z + \beta u/f. \tag{10.6}$$

Charney *et al.* ignored the β-term, which can be shown by scaling arguments to be small. They then expressed the advection term as a Jacobian:

$$\mathbf{V} \cdot \nabla \alpha = -\frac{g}{f}\frac{\partial z}{\partial y}\frac{\partial \alpha}{\partial x} + \frac{g}{f}\frac{\partial z}{\partial x}\frac{\partial \alpha}{\partial y} = -\frac{g}{f}J(\alpha, z). \tag{10.7}$$

Now using (10.6) and (10.7) in (10.5), they arrived at

$$\frac{\partial}{\partial t}(\nabla^2 z) = J\left(\frac{g}{f}\nabla^2 z + f, z\right). \tag{10.8}$$

This was taken as their basic equation (Eq. (8) in Charney *et al.*, 1950). It is interesting to observe that, had they chosen the stream-function rather than the geopotential as the dependent variable, they could have used the equation

$$\frac{\partial}{\partial t}(\nabla^2 \psi) = J(\nabla^2 \psi + f, \psi), \tag{10.9}$$

thereby avoiding the neglect of the β-term in (10.6). The boundary conditions required to solve (10.8) were investigated. It transpires that to determine the motion it is necessary and sufficient to specify z on the whole boundary and ζ over that part where the flow is inward. (The appropriate boundary conditions for (10.9) are ψ, or the normal velocity component, everywhere and ζ at inflow points.)

The vorticity equation was transformed to a polar stereographic projection; this introduces a map-factor, which we will disregard here. Initial data were taken from the manual 500 hPa analysis of the US Weather Bureau, discretised to a grid of 19×16 points with a grid interval corresponding to 8 degrees longitude at 45°N (736 km at the North Pole and 494 km at 20°N). Centred spatial finite differences and a leapfrog time-scheme were used. The boundary conditions were held constant throughout each 24-hour integration. Eq. (10.8) is equivalent to the system

$$\xi = \nabla^2 z \tag{10.10}$$

$$\frac{\partial \xi}{\partial t} = J\left(\frac{g}{f}\xi + f, z\right) \tag{10.11}$$

$$\nabla^2 \frac{\partial z}{\partial t} = \frac{\partial \xi}{\partial t}. \tag{10.12}$$

Given the geopotential height, ξ follows immediately from (10.10). The tendency of ξ is then given by (10.11). Next, the Poisson equation (10.12) is solved, with homogeneous boundary conditions, for the tendency of z, after which z and ξ are updated to the next time level. This cycle may then be repeated as often as required. Time-steps of one, two and three hours were all tried; with such a coarse spatial grid, even the longest time-step produced stable integrations.

The solution of the Poisson equation (10.12) was calculated by a Fourier transform method devised by von Neumann. This direct method was more suited to the ENIAC than an iterative relaxation method such as that of Richardson. Consider the Poisson equation $\nabla^2 \varphi = F$ with φ vanishing on the boundary of a rectangular region with grid

$$x_m = x_0 + (m/M)L_x \quad m = 0, 1, \ldots, M$$
$$y_n = y_0 + (n/N)L_y \quad n = 0, 1, \ldots, N$$

with $L_x = M\Delta s$ and $L_y = N\Delta s$. The standard five-point discretisation of the Laplacian is

$$(\nabla^2 \varphi)_{mn} = (\varphi_{m+1,n} + \varphi_{m-1,n} + \varphi_{m,n+1} + \varphi_{m,n-1} - 4\varphi_{m,n}) = \Delta s^2 F_{mn},$$

where Δs is the spatial interval. If φ is expanded in a double Fourier series

$$\varphi_{mn} = \sum_{k=1}^{M-1} \sum_{\ell=1}^{N-1} \tilde{\varphi}_{kl} \sin \frac{km\pi}{M} \sin \frac{\ell n\pi}{N},$$

the Laplacian can be applied separately to each term

$$(\nabla^2 \varphi)_{mn} = \sum_{k=1}^{M-1} \sum_{\ell=1}^{N-1} \left\{ -\frac{4}{\Delta s^2} \left(\sin^2 \frac{k\pi}{2M} + \sin^2 \frac{\ell\pi}{2N} \right) \right\} \tilde{\varphi}_{kl} \sin \frac{km\pi}{M} \sin \frac{\ell n\pi}{N}.$$

We can equate this term by term to the expansion of F to deduce $\tilde{\varphi}_{mn}$ and then compute the inverse transform to get φ:

$$\varphi_{mn} = -\frac{\Delta s^2}{MN} \sum_{i=1}^{M-1} \sum_{j=1}^{N-1} \sum_{k=1}^{M-1} \sum_{\ell=1}^{N-1} \left(\sin^2 \frac{k\pi}{2M} + \sin^2 \frac{\ell\pi}{2N} \right)^{-1}$$

$$\times F_{ij} \sin \frac{ik\pi}{M} \sin \frac{j\ell\pi}{N} \sin \frac{km\pi}{M} \sin \frac{\ell n\pi}{N}. \tag{10.13}$$

This is the expansion required to solve (10.12): if we replace F_{ij} by $(\partial \xi / \partial t)_{ij}$, then $\varphi_{mn} = (\partial z / \partial t)_{mn}$.

The data handling and computing operations involved for each time step are shown in Fig. 10.3 (from Platzman, 1979). Each row indicates a program specification by setting upwards of 5000 switches (column 1), a computation (column 2), the punching of output cards (column 3) and manipulation of cards on off-line equipment (column 4). Fourteen punch-card operations were required for each time step as the internal memory of ENIAC was limited to ten registers. The first row of Fig. 10.3 represents a step forward in time. The next depicts the computation of the Jacobian. Then follow four Fourier transforms, corresponding to the four-fold summation in (10.13). The final row indicates housekeeping computations and manipulations in preparation for the next step.

The computed forecast

The story of the mission to Aberdeen was colourfully told by Platzman (1979) in his Victor Starr Memorial Lecture:

On the first Sunday of March, 1950 an eager band of five meteorologists arrived in Aberdeen, Maryland, to play their roles in a remarkable exploit. On a contracted time scale the groundwork for this event had been laid in Princeton in a mere two to three years, but in another sense what took place was the enactment of a vision foretold by L. F. Richardson ...

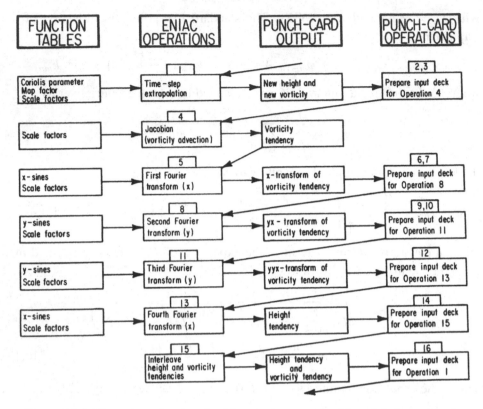

Figure 10.3 Flow chart showing the 16 operations required for each time step of the ENIAC forecast. (From Platzman (1979), © Amer. Met. Soc.)

[about 40] years before. The proceedings in Aberdeen began at 12 p.m. Sunday, March 5, 1950 and continued 24 hours a day for 33 days and nights, with only brief interruptions. The script for this lengthy performance was written by John von Neumann and by Jule Charney, who also was one of the five actors on the scene. The other players at Aberdeen were Ragnar Fjørtoft, John Freeman, [George Platzman and] Joseph Smagorinsky...

The trials and tribulations of this intrepid troupe were described in the lecture. There were the usual blunders familiar to programmers. The difficulties were exacerbated by the primitive machine language, the requirement to set numerous switches manually, the assignment of scale-factors necessitated by the fixed-point nature of ENIAC, and the tedious and intricate card-deck operations (about 100 000 cards were punched during the month). But, despite these difficulties, the expedition ended in triumph. Four 24-hour forecasts were made, and the results clearly indicated that the large-scale features of the mid-tropospheric flow could be forecast barotropically with a reasonable resemblance to reality. Each 24-hour integration took about 24 hours of computation; that is, the team were just able to keep pace

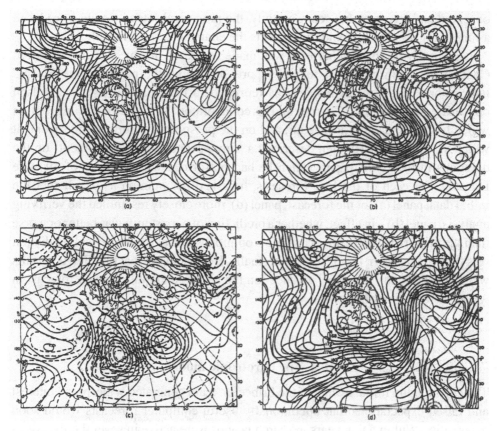

Figure 10.4 The ENIAC forecast starting at 0300 UTC, 30 January 1949. (a) Analysis of 500 hPa geopotential (thick lines) and absolute vorticity (thin lines). (b) Corresponding analysis valid 24 hours later. (c) Observed height changes (solid lines) and predicted changes (dashed lines). (d) Forecast height and vorticity. (Charney, 1951 © Amer. Met. Soc.)

with the weather. Much of the time was consumed by punch-card operations and manipulations. They estimated that when the IAS computer was ready the total elapsed time for a one-day forecast would be reduced to half an hour, 'so that one has reason to hope that Richardson's dream... of advancing the computation faster than the weather may soon be realised' (Charney *et al.*, 1950, p. 245).

The forecast starting at 0300 UTC, 30 January 1949 is shown in Fig. 10.4. Panel (a) is the analysis of 500 hPa geopotential (thick lines) and absolute vorticity (thin lines). Panel (b) is the corresponding analysis 24 hours later. Panel (c) shows the observed height changes (solid line) and predicted changes (dashed line). Finally, the forecast height and vorticity are shown in panel (d). It can be seen from panel (c) that there is a considerable resemblance between the general features of the observed and forecast changes. Certainly, the numerical integration has produced a forecast that is realistic and that reflects the changes in the atmosphere.

It is gratifying that Richardson was made aware of the success in Princeton; Charney sent him copies of several reports, including the paper on the ENIAC integrations. His letter of response is reprinted in Platzman (1968). Richardson opened by congratulating Charney and his collaborators 'on the remarkable progress which has been made in Princeton; and on the prospects for further improvement which you indicate'. He then described a 'tiny psychological experiment' on the diagrams in the *Tellus* paper, which he had performed with the help of his wife Dorothy. For each of the four forecasts, he asked her opinion as to whether the initial data, panel (a), or the forecast, panel (d), more closely resembled the verifying analysis, panel (b) – in effect, whether a prediction of persistence was better or worse than the numerical prediction. His wife's opinion was that the numerical prediction was on average better, though only marginally. He concluded that the ENIAC results were 'an enormous scientific advance' on the single, and quite wrong, forecast in which his own work had ended.

10.4 The barotropic model

The encouraging initial results of the Princeton team generated widespread interest and raised expectations that operationally useful computer forecasts would soon be a reality. Within a few years, research groups were active in several universities and national weather services. The striking success of the barotropic forecasts had come as a surprise to everyone. The barotropic vorticity equation simply states that the absolute vorticity of a fluid parcel is constant along its trajectory; thus, the relative vorticity ζ can change only as a result of a change in planetary vorticity f, which occurs when the parcel moves from one latitude to another. The dynamical importance of this β-effect had been clearly shown by Rossby (1939) in his linear study, and constant-absolute-vorticity (or CAV) trajectories were used in operational forecasting for some years, although with indifferent results. Clearly, the equation embodied the essence of mid-latitude wave dynamics but seemed too idealised to have any potential for operational use. The richness and power encapsulated in its nonlinear advection were greatly underestimated.

Not everyone was convinced that the barotropic equation was useful for forecasting. The attitude in some quarters to its use for prediction seems to have been little short of antagonistic. At a discussion meeting of the Royal Meteorological Society in January 1951, several scientists expressed strong reservations about it. In his opening remarks, Sutcliffe (1951) reviewed the application of the Rossby formula to stationary waves with a somewhat reluctant acknowledgement:

Although the connection between nondivergent motion of a barotropic fluid and atmospheric flow may seem far-fetched, the correspondence between the computed stationary wavelengths and those of the observed quasi-stationary long waves in the westerlies is found to be so good that some element of reality in the model must be suspected.

Bushby reported that tests of Charney and Eliassen's (1949) formula for predicting 500 hPa height changes showed some success but insufficient for its use in operations. He then discussed two experimental forecasts using the barotropic equation, for which the results had been discouraging 'probably due to the fact that baroclinity has been neglected'. Finally, he described the application of Sutcliffe's (1947) development theory, showing that it had some limited success in a particular case. He concluded that much further research was needed before numerical forecasting methods could be introduced on a routine basis. Richard Scorer expressed his scepticism about the barotropic model more vehemently, dismissing it as 'quite useless':

Even supposing that wave theory did describe the actual motion of moving disturbances in the atmosphere there is nothing at all to be gained by applying formulae derived for a barotropic model to obtain a forecast because all it can do is move on disturbances at a constant velocity, and can therefore give no better result than linear extrapolation and is much more trouble. *(Sutcliffe et al., 1951)*

Sutcliffe reiterated his doubts in his concluding remarks, saying that 'when a tolerably satisfactory solution to the three-dimensional problem emerges it will derive little or nothing from the barotropic model – which is literally sterile'. These meteorologists clearly had no confidence in the utility of the single-level approach.

The 1954 Presidential Address to the Royal Meteorological Society, *The Development of Meteorology as an Exact Science*, was delivered by the Director of the Met Office, Sir Graham Sutton. After briefly considering the methods first described in Richardson's 'strange but stimulating book', he expressed the view that automated forecasts of the weather were unlikely in the foreseeable future:

I think that today few meteorologists would admit to a belief in the possibility (let alone the likelihood) of Richardson's dream coming true. My own view, for what it is worth, is definitely against it. *(Sutton, 1954)*

He went on to describe the encouraging results that had been obtained with the Sawyer–Bushby model – making no reference to the activities at Princeton – but stressed that this work was not an attempt to produce a forecast of the weather. The prevalent view in Britain at that time was that while numerical methods had immediate application to dynamical research their use in practical forecasting was very remote. This cautious view may well be linked to the notoriously erratic nature of the weather in the vicinity of the British Isles and the paucity of data upstream over the Atlantic Ocean.

A more sweeping objection to the work at Princeton was raised by Norbert Wiener who, according to Charney, viewed the whole project with scorn, saying that the meteorological group were 'trying to mislead the whole world in[to] thinking that one could make weather predictions as a deterministic problem' (*Recollections*, p. 57). Charney felt that, in some fundamental way, Wiener was right and that he had anticipated the difficulty due to the unpredictability of the atmosphere, which was first considered in detail by Thompson (1957) and elucidated in a simple context by Lorenz (1963).[1] It is now generally accepted that there is indeed an inherent limit to the useful range of deterministic forecasts but in relation to short-range forecasting Wiener's view was unnecessarily gloomy.

Despite various dissenting views, evidence rapidly accumulated that even the rudimentary barotropic model was capable of producing forecasts comparable in accuracy to those produced by conventional manual means. In an interview in 1988, Ragnar Fjørtoft described how the success of the ENIAC forecasts 'had a rather electrifying effect on the world meteorological community'. When he returned to Norway in 1951, Fjørtoft was without access to computing machinery. Anxious to exploit the potential of the numerical methods for routine forecasting, he developed a graphical method of integrating the barotropic vorticity equation (Fjørtoft, 1952). His idea was to advect the absolute vorticity, using a smoothed velocity field that allowed a long time-step, and to solve a Helmholtz equation for the stream-function. All operations were performed graphically, using maps drawn on tracing paper. A 24-hour forecast could be calculated in a single time-step, the whole process taking less than three hours. The Princeton forecasts were re-done by the graphical method, and the results were of comparable accuracy to those using ENIAC. Fjørtoft's method was used for a time in 1952 and 1953 in several US Air Force forecast centres. Cressman (1996) regarded it as 'the first known operational use of numerical weather prediction'. Although this manual method was soon superseded by computer forecasts, it is historically important in that it links back to the graphical methods first proposed by Vilhelm Bjerknes and also forward to current methods: Fjørtoft seems to have been the first to employ the Lagrangian advection method which is so popular today.

The first computer integrations that were truly predictions, based on recent observations and available in time for operational use, were made in Stockholm in November 1954 (Persson, 2005a). Rossby had returned to Sweden in 1947 but maintained close links with the Princeton team, sending two of his students, Roy Berggren and Bert Bolin to the Institute for Advanced Studies. He was so impressed by the success of the ENIAC experiments that he set up a collaborative project between the newly founded International Meteorological Institute at Stockholm

---

[1] The predictability of the atmosphere is discussed in §11.6 below.

University and the Royal Swedish Air Force, to carry the work over to an operational context. A barotropic model was integrated out to 72 hours on a Swedish computer, BESK, similar to the Princeton machine. The results of these trials were very encouraging: the average correlation between computed and observed 24-hour changes was significantly better than that obtained with conventional methods. When an automatic analysis scheme had been developed (Bergthorsson and Döös, 1955), regular operations were initiated in 1956. Col. Herrlin (RSAF), reporting these results at a Symposium on NWP in Frankfurt in May 1956, described the situation thus:

For the last 10 years, or so, the progress made in the technique of 1–2 days forecasts has been very small. The development appears to have become almost stagnant. I believe we have literally squeezed the conventional technique dry. . . . When therefore Prof. Rossby told me that he was convinced that a practicable program for numerical forecasts of the 500 mb surface now was available, we within the Swedish Military Weather Service were eager to develop and test this system in our general routine. As has been shown, our experiences have been most favourable and I feel convinced that this is only the modest beginning of a new era ... comparable with the era created by the Norwegian school in the nineteen twenties.   *(Herrlin, 1956)*

The Swedish work represented the inauguration of the era of operational objective forecasting based on scientific principles. It is reviewed by Persson (2005a) and by Wiin-Nielsen (1997). The humble barotropic vorticity equation continued to provide useful guidance for almost a decade.

10.5 Multi-level models

On Thanksgiving Day 1950 a severe storm caused extensive damage along the east coast of the United States (Smith, 1950). The prediction of rapid cyclogenesis events like this had long been considered as the central problem in synoptic meteorology. The simple barotropic model is incapable of representing such explosive deepening, as it does not allow for the energy transformations that are crucial for such developments. Charney (1947) and Eady (1949) had elucidated the role of baroclinic instability in cyclogenesis. It was clear that the prediction of this phenomenon required a numerical model that accounted for vertical variations and allowed for conversion of available potential energy to kinetic energy. Several baroclinic models were developed in the few years after the ENIAC forecast (Árnason, 1952; Charney and Phillips, 1953; Eady, 1952; Eliassen, 1952; Phillips, 1951; Sawyer and Bushby, 1953). They were all based on the quasi-geostrophic system of equations. The Princeton team studied the Thanksgiving storm using two- and three-level models. After some tuning, they found that the cyclogenesis could be reasonably well simulated. Thus, it appeared that the central problem of operational forecasting had been cracked.

These results were instrumental in persuading the Air Weather Service, the Naval Weather Service and the US Weather Bureau to combine forces and establish the Joint Numerical Weather Prediction Unit (JNWPU). The unit came into being in July 1954 with George Cressman as Director, Joseph Smagorinsky as head of the operational section, Philip Thompson as head of the development section and Art Bedient as head of the computer section, and regular operations began about one year later. The first experimental model was the three-level model of Charney and Phillips. However, it transpired that the success of the Thanksgiving forecast had been something of a fluke. Shuman (1989) reports that the multi-level models were consistently worse than the simple barotropic equation. As a result, the single-level model was used for operations from 1958. There was an immediate improvement in forecast skill over subjective methods, but this had come only after intensive efforts to remove some deficiencies that had been detected in both the barotropic and multi-level models. A spurious anticyclogenesis problem was solved by replacing geostrophic initial winds by non-divergent winds derived using the balance equation (Shuman, 1957), and spurious retrogression of the longest waves was suppressed by allowing for the effects of divergence (Cressman, 1958). In fact, both these problems had also been noted by the Stockholm group and similar solutions had already been devised by Bolin (1956).

The grid resolution of the first operational model at JNWPU was 381 km (Wiin-Nielsen, 1997). This number, which may appear strange, was chosen for a practical reason. If the map-scale is $1 : M \times 10^6$ (one to M million) and the grid size in kilometres is Δ, then the distance d between grid points on the map is $d = \Delta/M$ millimetres. However, the line-printers used for displaying the zebra-chart plots were designed in imperial units. The distance *in inches* between grid points is $d = \Delta/(25.4 \times M)$. Thus, $\Delta = 381$ km gave $d = 1.5''$ on a one-to-ten-million map. There were ten print characters per inch and six lines per inch; the location of grid points at print points removed the need for numerical interpolation. Thus, model resolutions of 127 km, 254 km and 381 km were common for early models.[2] Indeed, these resolutions continued to be used even after line-printers were replaced by high-resolution plotters. The grid resolution of the early models had to be very coarse to ensure adequate geographical coverage for a one-day forecast. In his monograph *Dispersion Processes in Large-scale Weather Prediction*, Phillips wrote:

If Charney and his collaborators had chosen too small an area in which to make their computations, the first modern attempt at numerical weather prediction would have been severely degraded by the spread of errors from outside the small forecast area.

(Phillips, 1990)

---

[2] The grid unit of 381 km was popularly called a *Bedient*, after Art Bedient who first thought up the idea.

Phillips observed that it was fortunate that Charney had applied group velocity concepts so that a reasonable decision could be made about the minimum forecast area and that, had the region been too small, the ENIAC results 'might have been as discouraging as was Richardson's attempt 30 years earlier'. The additional computational demands of multi-level models meant that the geographical coverage was more limited. This increased the risk of corruption of the forecast by errors propagating from the lateral boundaries, where the variables retained their initial values. Persson (2005c) argues that this was the reason why the early baroclinic model results in Britain were unsatisfactory.

The size and scope of the International Symposium on Numerical Weather Prediction that was organised in Frankfurt in 1956 indicates the state of play at that time. There were over 50 participants from USA, from Japan and from 11 European countries. Some 27 contributions are contained in the report (DWD, 1956), including several from the German pioneers, Hinkelmann, Hollman, Edelmann and Wippermann (known colloquially as *Die Viermännergruppe* or simply *Die Männer*, the men). Although operational NWP was not introduced in Germany until 1966, there were significant theoretical developments from much earlier. As there was no access to computers, the first integrations were done manually, using a filtered three-level baroclinic model. In an interview (Taba, 1988) Winardt Edelmann tells the story of some early work in the autumn of 1952. First, the geopotential analysis at three levels was prepared by the synopticians.

Then the vorticity and even the Jacobians of the quasi-geostrophic model had to be evaluated by graphical addition and subtraction and more elaborate methods, producing a whole lot of maps; that took several days and was not very precise. Then a square grid was placed over the Jacobian maps and values interpolated for each grid point. . . . [A]fter several weeks we had figures we assumed to be the solution to the elliptic equation. The tendency was converted back to a map and graphically added to the initial field, giving us a forecast for 12 hours. Then the entire operation had to be repeated to give a 24-hour prediction. The result did not look totally unreasonable.

In 1955, George Platzman conducted a worldwide survey to assess the level of activity in numerical weather prediction. A report on the results was distributed the following year (Birchfield, 1956). Numerical methods were already under active investigation in USA, Britain, Sweden, Germany and Belgium. Objective graphical techniques were in use or under study in USA, Japan, Ireland and New Zealand. Preliminary activities or immediate plans were reported by Canada, Finland, Israel, Norway and South Africa. Of course, this survey could not be complete. There were also activities at an early stage in Australia, France, Russia and elsewhere. Persson (2005b, 2005c) has reviewed early operational numerical weather prediction outside the USA, paying particular attention to developments in Britain.

By 1960, numerical prediction models based on the filtered equations were either operational or under investigation at several national weather centres. Baroclinic models were being developed, but they did not yield dramatic improvements over the barotropic models. There were inherent shortcomings due to the approximations implicit in the filtered equations. The assumption of geostrophy gives rise to errors associated with the variation of the Coriolis parameter with latitude. An assumption of quasi-nondivergence would have circumvented this problem to some extend. In a review of early numerical prediction, Phillips (2000) wrote: 'I believe that baroclinic quasi-geostrophic models might have been more productive... if they had used a stream function in place of the geopotential'. While this may be true, there were other severe limitations with the filtered equations, one of which was their inapplicability in the tropics. To overcome these difficulties, a return to the method originally employed by Richardson was necessary.

10.6 Primitive equation models

The limitations of the filtered equations were recognised at an early stage. In a forward-looking synopsis in the *Compendium of Meteorology*, Jule Charney wrote:

The outlook for numerical forecasting would indeed be dismal if the quasi-geostrophic approximation represented the upper limit of attainable accuracy, for it is known that it applies only indifferently, if at all, to many of the small-scale but meteorologically significant motions.   *(Charney, 1951)*

Charney discussed some integrations that he had performed with John Freeman using a linear barotropic primitive equation model. The computed motion was found to consist of two superimposed parts, a Rossby motion and gravity wave motion of much smaller amplitude:

In a manner of speaking, the gravity waves created by the slight unbalance served the telegraphic function of informing one part of the atmosphere what the other part was doing, without themselves influencing the motion to any appreciable extent.   *(Charney, 1951)*

He considered the prospects for using the primitive equations, and argued that if geostrophic initial winds were used, the gravity waves would be acceptably small. Within a year or so of arriving at Princeton, Charney had realised that it would be possible to integrate the primitive equations provided the CFL stability criterion were satisfied. There would be gravity wave oscillations in the solution but their amplitude would remain bounded: 'It would give you an embroidered tendency field which would be essentially correct. In other words, the primitive equations would be quite possible' (Platzman, 1990 [*Recollections*] p. 38). This indicates a volte-face from the view that he had formed when he first studied Richardson's book: 'I thought that maybe... the primitive equations were just not appropriate' (*Recollections*, p. 39). In his Compendium article, Charney outlined a scheme for solving the primitive equations, but cautioned that, in the last analysis, the feasibility of using them would be determined only by actual numerical integrations.

In a letter to Platzman, Charney wrote that he had been 'greatly encouraged' by Richardson's generous remarks on the first numerical forecasts on ENIAC, and had subsequently sent him a report on the baroclinic integrations:

Sad to say... it arrived five days after his death. I wish now I had earlier sent him an article I wrote for the *Compendium of Meteorology* in which, at the end, I came to the conclusion that his approach was perfectly feasible despite the initial-value problem providing only that one was careful to satisfy the condition of computational stability. Of course, his work needs no vindication from me.
(Platzman, 1968)

Research with the primitive equations began at NMC (now NCEP) in 1959. A six-level primitive equation model was introduced into operations in June 1966, running on a CDC 6600 (Shuman and Hovermale, 1968). There was an immediate improvement in skill: the S_1 score (Teweles and Wobus, 1954) for the 500 hPa one-day forecast was improved by about five points. Platzman (1967) made a detailed comparison between the Shuman–Hovermale model and Richardson's model. While there were significant differences, the similarities were more striking. Even the horizontal and vertical resolutions of the two models were quite comparable. This is all the more surprising as the NMC model was not designed by consciously following Richardson's line of development but had evolved from the earlier modelling work at Princeton, together with Eliassen's (1949) formulation of the equations in isobaric co-ordinates.

Karl-Heinz Hinkelmann had been convinced from the outset that the best approach was to use the primitive equations. He knew that they would simulate the atmospheric dynamics and energetics more realistically than the filtered equations. Moreover, he felt certain, from his studies of noise, that high frequency oscillations could be controlled by appropriate initialisation. His 1951 paper 'The mechanism of meteorological noise' was the first systematic attempt to tackle the issue of suitable initial conditions. In it, he argued that geostrophic winds would yield a forecast substantially free from high frequency noise. Furthermore, the extra computation, necessitated by shorter time-steps, to integrate the primitive equations would be partially offset by the simpler algorithms, which involved no iterative solution of elliptic equations. His first application of the primitive equations was a success, producing good simulation of development, occlusion and frontal structure. In an interview for the WMO Bulletin he said:

On my first attempt, using idealized initial data, I got a most encouraging result which reproduced new developments, occlusions and even the kinks in the isobars along a front.
(Taba, 1988)

Soon after they had done that first run with the primitive equations, Hinkelmann and his team visited Smagorinsky in Washington, D.C.

After seeing our results, he [Smagorinsky] said that we had done a fine job, but added that his group also had good results with the primitive equations and intended to use them exclusively from that time on. So in fact our independent research efforts had both led to the same conclusion. I consider that the change from quasi-geostrophic models to primitive equations was a very important step in simulating atmospheric processes.

Routine numerical forecasting was introduced in the Deutscher Wetterdienst in 1966; according to Reiser (2000), this was the first ever use of the primitive equations in an operational setting. In an interview in November 1987, André Robert was asked if improvements in numerical weather prediction had been gradual or if he knew of a particular change of model that produced a dramatic improvement. He replied that with the first primitive equation model in Washington there were drastic improvements, and the decision was made immediately to abandon filtered models for operational forecasting (Lin *et al.*, 1997).

The first primitive equation models (Smagorinsky, 1958; Hinkelmann, 1959) were adiabatic, with dry physics. The introduction of moisture brought additional serious problems. Conditional instability leads to the rapid development of small-scale convective systems, called grid-point storms, and larger synoptic-scale depressions and tropical cyclones are starved of the energy necessary for their growth. To rectify this problem, convective instability must be reduced throughout the unstable layer. In the early diabatic models this was achieved by a process called *moist convective adjustment*, which suppresses gravitational instability. Kasahara (2000) has written an interesting history of the development of cumulus parameterisations for NWP in which he argues that cumulus schemes were a critical factor in enabling stable time integrations of primitive equation models with moist physical processes.

We have seen that the view in Britain was that single-level models were unequal to the task of forecasting. As a result, barotropic models were never used for forecasting at the UK Met Office and, partly for this reason, the first operational model (Bushby and Whitelam, 1961) was not in place until the end of 1965, more than ten years after operational NWP commenced at JNWPU in Washington. In 1972 a ten-level primitive equation model (Bushby and Timpson, 1967) was introduced. This model incorporated a sophisticated parameterisation of physical processes including heat, moisture and momentum through the bottom boundary, topographic forcing, sub-grid-scale convection and lateral diffusion. Useful forecasts of precipitation were produced. An example of one such forecast is shown in Fig. 10.5: the top panel shows the estimated total rainfall for 06–18 UTC on 1 December 1961, based on weather reports; the bottom panel shows the forecast total rain for the same period. The maximum over Southern Ireland and Britain is reasonably

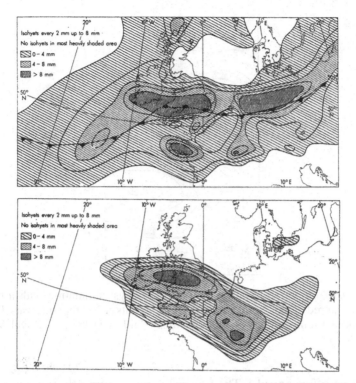

Figure 10.5 Top panel: estimated total rainfall for 06–18 UTC on 1 December 1961, based on weather reports. Bottom panel: forecast total rain for the same period based on the Bushby–Timpson model. (From Benwell *et al.* (1971), © HMSO.)

well predicted. The maximum over Northern Germany is poorly reflected in the forecast. The results indicate that the model was capable of producing a realistic rainfall forecast.

Despite the initial hesitancy in Britain to give credance to computer forecasts, the following account appeared in a popular exposition of numerical forecasting in the late 1970s:

It was the Meteorological Office of Great Britain, Richardson's own country, that was the first weather-forecasting service of any country in the world to acquire a computer, the IBM 360/195, big enough to carry out regular weather prediction with a highly detailed and refined process along ... [Richardson's] lines. *(Lighthill, 1978)*

Such hyperbole is superfluous: the Met Office has continued to hold a leading position in the ongoing development of numerical weather prediction. A 'Unified Model', which may be configured for global, regional and mesoscale forecasting and as a general circulation model for climate studies, was introduced in 1993 (Cullen, 1993) and continues to undergo development (see

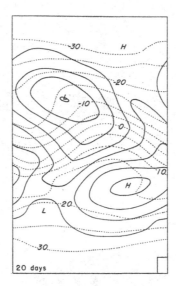

Figure 10.6 Configuration of the flow after 20 days' simulation with a simple, two-level filtered model. Solid lines: 1000 hPa heights at 200 foot intervals. Dashed lines: 500 hPa temperatures at 5°C intervals. (Phillips (1956), © Royal Meteor. Soc.)

http://www.metoffice.gov.uk). The forecast version, currently running on an NEC SX-8/128M16 computer system with a total of 128 processing elements, is now the basis of operational weather prediction at the Met Office. The model is also the primary resource for climate modelling at the Hadley Centre, a leading centre for the study of climate and climate change.

10.7 General circulation models and climate modelling

Norman Phillips carried out the first long-range simulation of the general circulation of the atmosphere. He used a two-level quasi-geostrophic model on a beta-plane channel with rudimentary physics. The computation, done on the IAS computer (MANIAC I), used a spatial grid of 16×17 points, and the simulation was for a period of about one month. Starting from a zonal flow with small random perturbations, a wave disturbance with wavelength of 6000 km developed. It had the characteristic westward tilt with height of a developing baroclinic wave, and moved easward at about $20 \, \mathrm{m \, s^{-1}}$. Figure 10.6 shows the configuration of the flow after 20 days' simulation. Phillips examined the energy exchanges of the developing wave and found good qualitative agreement with observations of baroclinic systems in the atmosphere. He also examined the mean meridional flow, and found circulations corresponding to the Hadley, Ferrel and Polar cells:

We see the appearance of a definite three-celled circulation, with an indirect cell in middle latitudes and two somewhat weaker cells to the north and south. This is a characteristic feature of . . . unstable baroclinic waves.   *(Phillips, 1956, p. 144)*

John Lewis has re-examined Phillips' experiment and the circumstances that led up to it (Lewis, 1998). Phillips presented this work to a meeting of the Royal Meteorological Society, where he was the first recipient of the Napier Shaw Prize. The leading British dynamicist Eric Eady said, in the discussion following the presentation, 'I think Dr Phillips has presented a really brilliant paper which deserves detailed study from many different aspects.' Von Neumann was also hugely impressed by Phillips' work, and arranged a conference at Princeton University in October 1955, *Application of Numerical Integration Techniques to the Problem of the General Circulation*, to consider its implications. The work had a galvanising effect on the meteorological community. Within ten years, there were several major research groups modelling the general circulation of the atmosphere, the leading ones being at the Geophysical Fluid Dynamics Laboratory (GFDL), the National Center for Atmospheric Research (NCAR) and the Met Office.

Following Phillips' seminal work, several general circulation models (GCMs) were developed. One early model of particular interest is that developed at NCAR by Kasahara and Washington (1967). A distinguishing feature of this model was the use of height as the vertical co-ordinate (most models used pressure p or normalised pressure σ). The vertical velocity was derived using Richardson's Equation; indeed, the dynamical core of this model was very similar to that employed by Richardson. The Kasahara–Washington model was a simple two-layer model with a 5° horizontal resolution. It was the first in a continuing series of climate models. Various physical processes such as solar heating, terrestrial radiation, convection and small-scale turbulence were included in these models. The Community Atmosphere Model (CAM 3.0) is the latest in the series. CAM also serves as the atmospheric component of the Community Climate System Model (CCSM) a 'fully-coupled, global climate model that provides state-of-the-art computer simulations of the Earth's past, present, and future climate states.' Thanks to enlightened American policy on freedom of information, these models are available to the weather and climate research community throughout the world, and can be downloaded from the NCAR website without cost (www.ncar.ucar.edu).

A declaration issued at the World Economic Forum in Davos, Switzerland in 2000 read: *Climate change is the greatest global challenge facing humankind in the twenty-first century.* There is no doubt that the study of climate change and its impacts is of enormous importance for our future. Global climate models are the best means we have of anticipating likely changes. The latest climate model (HadCM3) at the Hadley Centre is a coupled atmosphere–ocean general circulation model.

Many earlier coupled models needed a flux adjustment (additional artificial heat and moisture fluxes at the ocean surface) to produce good simulations. The higher ocean resolution of HadCM3 was a major factor in removing this requirement. To test its stability, HadCM3 has been run for over 1000 years simulated time and shows minimal drift in its surface climate. The atmospheric component of HadCM3 has 19 levels and a latitude/longitude resolution of $2.5° \times 3.75°$, with grid of 96×73 points covering the globe. The resolution is about 417×278 km at the equator. The physical parameterisation package of the model is very sophisticated. The radiative effects of minor greenhouse gases as well as CO_2, water vapour and ozone are explicitly represented. A parameterisation of background aerosol is included. The land surface scheme includes freezing and melting of soil moisture, surface run-off and soil drainage. The convective scheme includes explicit down-draughts. Orographic and gravity wave drag are modelled. Cloud water is an explicit variable in the large-scale precipitation and cloud scheme. The atmospheric component of the model allows the emission, transport, oxidation and deposition of sulphur compounds to be simulated interactively. The oceanic component of HadCM3 has 20 levels with a horizontal resolution of $1.25° \times 1.25°$ permitting important details in the oceanic current structure to be represented. The model is initialised directly from the observed ocean state at rest, with a suitable atmospheric and sea ice state. HadCM3 is being used for a wide range of climate studies which will form crucial inputs to the forthcoming Fourth Assessment Report (AR4) of the Intergovernmental Panel on Climate Change (IPPC), to be published in 2007.

The development of comprehensive models of the atmosphere is undoubtedly one of the finest achievements of meteorology in the twentieth century. Advanced models are under continuing refinement and extension, and are increasing in sophistication and comprehensiveness. They simulate not only the atmosphere and oceans but also a wide range of geophysical, chemical and biological processes and feedbacks. The models, now called *Earth System Models*, are applied to the eminently practical problem of weather prediction and also to the study of climate variability and mankind's impact on it.

11

Numerical weather prediction today

With the recent rapid development of high speed, large scale computing
devices, Richardson's work has ceased to be merely of academic interest
and has become of the greatest importance . . . for the solution of meteorological problems.   (*Charney, 1950,* p. 235)

11.1 Observational data

The initial data required to make a numerical forecast are derived ultimately from
direct observations of the atmosphere. At the time of Richardson's forecast, surface observations were made at regular times throughout the day at a network of
dedicated weather stations. However, the network was very uneven, with a concen-
tration of stations in densely inhabited areas. As Richardson wrote in the preface
of *WPNP*:

The present distribution of meteorological stations on the map has been governed by various
considerations: the stations have been outgrowths of existing astronomical or magnetic
observatories; they have adjoined the residence of some independent enthusiast, or of the
only skilled observer available in the district; they have been set out upon the confines of the
British Isles so as to include between them as much weather as possible; or they have been
connected with aerodromes in order to exchange information with airmen. On the map the
dots representing the positions of the stations look as if they had fallen from a pepperpot.
The nature of the atmosphere, as summarized in its chief differential equations, appears to
have been without influence upon the distribution.

Richardson's ideal solution – quite impractical – was for a weather station to be
located at the centre of each computational cell, as illustrated in the frontispiece of
his book (see Fig. 1.6 on p. 21).

The establishment and maintenance of an observational network requires close
international collaboration. Fortunately, co-operation between national meteorological services has traditionally been excellent. The International Meteorological

209

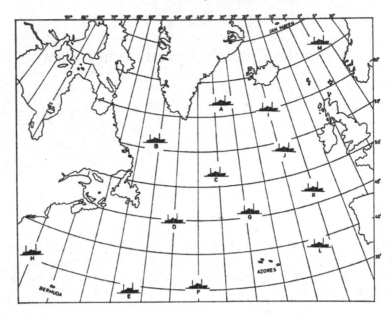

Figure 11.1 The PICAO array of 13 weather ships established following international conferences in Dublin (March 1946) and London (September 1946). (From
Petterssen (2001), © Amer. Met. Soc.)

Organization (IMO), an informal 'club' of directors of the national weather services
established in 1873, essentially collapsed during the hostilities of World War II. It
became clear during the war that forecasts for more than a few hours ahead required
comprehensive coverage over a wide geographical region. The forecasts for D-day
(6 June 1944) depended critically on an observation from Blacksod in the extreme
north-west of Ireland (Petterssen, 2001). There is no doubt that, had observations
for the Atlantic Ocean to the west of Ireland been available, they would have been
of inestimable value.[1]

It was not until 1951 that IMO was reconstituted as the World Meteorological
Organization (WMO), a specialised agency of the United Nations. In the meantime,
the requirements of transatlantic civil aviation had to be addressed. The Provisional
International Civil Aviation Organization (PICAO) was established, on the initiative
of the United States (it later became ICAO when the qualifier *Provisional* was
dropped following ratification by the member countries). At a conference in Dublin
in March 1946, a proposal was made to establish an array of 13 weather ships in
the North Atlantic (see Fig. 11.1). This system was put in place with remarkable
speed, the ships being funded by the United States and Canada (8), the United
Kingdom (2), France (1), Belgium and Holland (1) and Norway and Sweden (1).

---

[1] The results of a recent re-analysis of the weather on D-day are available on the ECMWF website
http://www.ecmwf.int/research/era/dday/

The PICAO ships made radiosonde observations twice a day at fixed locations. In addition, many merchant ships made observations of surface conditions at six-hourly intervals. The history of the PICAO array is described by Petterssen (2001) who remarked that developments in computer forecasts would certainly not have been possible without the upper-air observations provided by this network of ocean stations. Of course, the coverage provided by surface-based observing stations was far below the 'ideal' of an observation at each grid point. The arrival of weather satellites changed the situation dramatically.

Meteorological satellites

Sputnik I was launched in October 1957. While yielding little in the way of useful geophysical data, it provided a strong impetus to the United States, kicking off the space-race. Since the mid 1960s, the entire globe has been observed on a daily basis. *TIROS-1* was launched on April Fool's Day 1960 and immediately began sending back television pictures of the Earth's atmosphere, which generated great excitement. The value of these images for operational weather forecasting was soon recognised. There followed a long series of polar orbiting satellites carrying instrumentation of increasing sophistication and versatility. In addition to the cameras which provided visual and infra-red imagery, these satellites carried an array of radiometers that measured the radiation emanating from the atmosphere in a wide range of spectral bands. From these data, vertical profiles of temperature and moisture could be deduced. The first satellite temperature retrievals were made in 1969 using the satellite infra-red spectrometer (SIRS) instrument on *Nimbus-3*. The retrieval involved the inversion of the radiative transfer equation. Temperature profiles were derived using emissions from CO_2, and moisture content was then inferred from H_2O emissions. By 1980, *TIROS-N* was providing complete global coverage of vertical temperature and moisture profile data every 12 hours at a spacing of 250 km (Smith *et al.*, 1981).

The effort to extract useful information from satellite soundings has been one of the great struggles of the past 40 years, and there have been many heroic failures. The TOVS (*TIROS* Operational Vertical Sounder) data were introduced into NWP models at an early stage. However, the resulting improvements in forecast accuracy were very modest. There were several reasons for this: the radiance measurements had significant and systematic errors; the inversion process was inherently inaccurate; cloudy skies had a strong influence that was difficult to allow for. While the satellite data provided extensive geographical coverage, they were asynchronous, with observations at different times in different places. These characteristics posed substantial difficulties for conventional analysis systems. The successful exploitation of the satellite data had to await the development of modern data assimilation systems.

Complementing the polar orbiters, a geostationary satellite *SMS-1* was launched in 1974, positioned over the equator at 75°W. *SMS-2*, located at 135°W, followed one year later. Series of images from these satellites could be used to track cloud movements from which wind vectors at cloud level could be inferred. The evolution of mesoscale features could be observed by viewing the imagery in animation. This provided a strong impetus to the development of meteorology. Hurricanes that form over tropical oceanic regions are now spotted at an early stage of their evolution, and can be tracked with great accuracy. Time-lapse loops of cloud imagery have greatly improved the prediction of severe local storms. Today, a set of five geostationary satellites provides continuous global coverage, representing 'the equivalent of a reporting station every 1 km with visible data (every 4 km with infrared data)' (Purdom *et al.*, 1996). Such data coverage is beyond even Richardson's wildest imaginings.

The global observing system

The global observing system has changed considerably over the past 50 years. The synoptic surface observations from land stations and ships, and the soundings from radiosonde and pilot balloons have been the backbone of the system, but things are changing as satellite-borne instruments displace ground-based systems. The quality of the radiosonde measurements has improved, but the geographical coverage has declined. Satellite data have compensated for this decline. Many new types of observation have been introduced over the period. The schematic diagram reproduced in Fig. 11.2, from Uppala *et al.* (2005), shows how 1973 was a key year, with a significant increase in the number of available aircraft observations, the first observations from ocean buoys, and radiances from the vertical temperature profile radiometer (VTPR) instruments flown on the early National Oceanic and Atmospheric Administration (NOAA) series of operational polar-orbiting satellites.

The Global Weather Experiment, also known as FGGE, began in 1979 (FGGE was the First GARP Global Experiment; GARP was the Global Atmospheric Research Programme). There was a major enhancement of the observing system for this event. VTPR data were replaced by data from more sophisticated TOVS instruments on the new NOAA platforms, winds derived from geostationary satellites first became available and there was a large increase in the number of buoy and aircraft data. Ozone data also become available at about this time. After FGGE, observation density declined for a while, but recovered during the 1980s. The density of wind and temperature measurements from aircraft and winds from geostationary satellites increased substantially in the 1990s. Newer satellite instruments include SSM/I (Special Sensor Microwave Imager), the ERS altimeter for ocean-wave analysis (ERS is the ESA Remote Sensing Satellite; ESA is the European Space Agency),

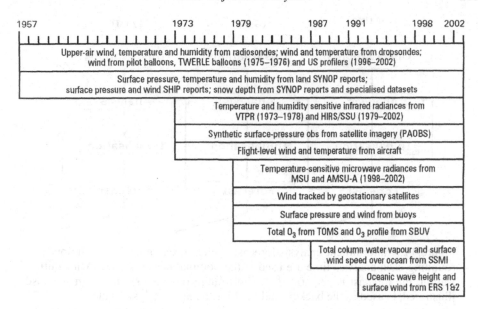

Figure 11.2 Development of the global observing system over the past 50 years. Conventional surface and upper-air data have been available throughout the period. Satellite radiance data have been available since 1973, and enhanced satellite data since 1979. (From Uppala *et al.* (2005), © ECMWF).

the ERS scatterometer (for surface winds) and AMSU-A (the Advanced Microwave Sounding Unit). Today's observing system provides coverage of the entire globe with high spatial density and high temporal resolution.

11.2 Objective analysis

Numerical weather prediction is an 'initial value problem': to integrate the equations of motion, we must specify the values of the dependent variables at an initial time. The numerical process then generates the values of these variables at later times. Richardson obtained his initial values in an ad hoc fashion. Using the few observations available to him, he deduced the required values by manual interpolation to his grid points. The initial data for the ENIAC integrations were also produced by hand, a labour-intensive and time-consuming process. Computed weather forecasts are of operational use only if they are available well ahead of the verifying time. To ensure this, it was essential to automate the process of preparing the initial fields. This process is called objective analysis. A comprehensive review of conventional objective analysis methods was provided by Daley (1991). There have been numerous conferences and workshops devoted to this topic and the many reports produced by the European Centre for Medium-Range Weather Forecasts are a particularly valuable resource.

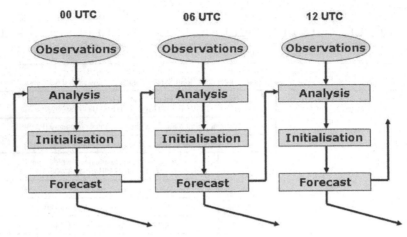

Figure 11.3 Intermittent analysis/forecast cycle. Observations in a window sur-rounding each analysis time are used at the nominal analysis time. After initial-isation, the forecast is performed. A short-range forecast, six hours in the case illustrated, is used as the background field for the next analysis cycle.

One of the earliest versions of objective analysis, developed by Panofsky (1949), used a high-order polynomial to derive a global approximation to the observational data. This was a purely empirical method, with no sound theoretical justifica-tion. Gilchrist and Cressman (1954) developed a local fitting method, using a bi-quadratic equation

$$z(x, y) = a_1 + a_2 x + a_3 y + a_4 x^2 + a_5 xy + a_6 y^2.$$

The coefficients $\{a_n\}$ were determined separately for each grid point by minimis-ing the mean square difference between the polynomial value and observed values close to the grid point. The method had a severe drawback: in data-sparse regions, there was no basis for the analysis. Following the practice of operational fore-casters, Bergthorsson and Döös (1955) introduced a 'first-guess' or background field comprised of climatological values, a short-range forecast or a combination of these. This field was then modified in the vicinity of observational data. An 'analysis cycle' of successive short-range forecasts and analyses was then feasible (see Fig. 11.3). This procedure enabled the transport of information from data-rich to data-poor regions through advection during the forecast stage. With the paucity of observations over ocean areas, this was a crucial development. The analysis of Bergthorsson and Döös (1955) was used in the early operational NWP system in Sweden. Analysis cycles with assimilation of observations every six hours are still in widespread use today.

Cressman (1959) devised an objective analysis procedure known as the suc-cessive correction method. The background field was modified repeatedly, using an empirical weighted adjustment, to concur with the observed values. The spatial scale of the adjustment was reduced with each iteration, so that synoptic scales were fitted first and smaller scale features in subsequent iterations. Brathseth (1986) later showed that, with an appropriate choice of the weighting functions, the successive correction method converges to the process of optimal interpolation, which we consider next.

Optimal interpolation

A method of optimising the use of information in the background field and in the observations was first proposed by Arnt Eliassen and developed in much more detail by Gandin (1963). This method, called optimal interpolation (OI), used the statistical properties of the forecast and observation errors to produce an analysis, which, in a precise statistical sense, is the best possible analysis. The OI analysis method was for several decades the most popular method of automatic analysis for numerical weather prediction. We will describe the method briefly now; for a fuller account, see Daley (1991) or Kalnay (2003).

The state of the atmosphere at a particular time may be given in terms of a vector $\mathbf{X}$ of the variables at all the model grid points, the *state vector*. We wish to deduce the best estimate $\mathbf{X}_A$ of $\mathbf{X}$ from the available data. Let us assume that a background field $\mathbf{X}_B$ is given; this is normally a short-range forecast referring to the analysis time. We are also given a vector $\mathbf{Y}$ of observations. Note that $\mathbf{Y}$ differs in character from $\mathbf{X}_B$, since the observations are irregularly distributed and the variables observed may not be the model variables. For example, satellites measure radiances, not temperatures. We express the vector of analysed values as

$$\mathbf{X}_A = \mathbf{X}_B + \mathbf{K}[\mathbf{Y} - \mathbf{H}(\mathbf{X}_B)] \tag{11.1}$$

where $\mathbf{H}$ is the 'observation operator', which converts the background field into first guess values of the observations. This operator includes spatial interpolation to observation points and transformation to observed variables based on physical laws. In general, $\mathbf{H}$ is nonlinear but, for simplicity, we will assume here that it is linear. Eq. (11.1) says that the analysis is obtained by adding to the background field a weighted sum of the difference between observed and background values. The matrix $\mathbf{K}$ of weights, the *gain matrix*, is chosen so as to minimise the root-mean-square analysis error. It turns out that $\mathbf{K}$ is given by

$$\mathbf{K} = \mathbf{B}\mathbf{H}^T(\mathbf{R} + \mathbf{H}\mathbf{B}\mathbf{H}^T)^{-1} \tag{11.2}$$

where $\mathbf{B}$ is the covariance matrix of background field errors and $\mathbf{R}$ the observational error covariance matrix (see, e.g. Kalnay, 2003). Moreover, the analysis error covariance is then given by

$$A = (I - KH)B.$$

In OI, the background and observation error matrices B and R and the observation operator H are modelled in an empirical manner and are assumed to be constant in time. (In Kalman filter analysis, which we will not discuss, B is *predicted* from the previous analysis time.) In practical implementations of OI, numerous simplifications must be introduced that degrade the precision of the analysis. Thus, the name statistical interpolation is probably more apposite than optimal interpolation.

Variational assimilation

An alternative approach to data assimilation is to find the analysis field that minimises a *cost function*. This is called variational assimilation and it is equivalent to the statistical technique known as the maximum likelihood estimate, subject to the assumption of Gaussian errors. When applied at a specific time, the method is called three-dimensional variational assimilation or, for short, 3D-Var. When the time dimension is also taken into account, we have 4D-Var. The cost function for 3D-Var may be defined as

$$J = \underbrace{(X - X_B)^T B^{-1}(X - X_B)}_{J_B} + \underbrace{(Y - HX)^T R^{-1}(Y - HX)}_{J_O} \qquad (11.3)$$

which is the sum of two components. The term J_B represents the distance between the analysis X and the background field X_B weighted by the background error B. The term J_O represents the distance between the analysis and the observed values Y, weighted by the observation error covariance R. To keep things simple, we again assume a linear observation operator H. The minimum of J is attained at $X = X_A$ where

$$\nabla_X J = 0.$$

that is, where the gradient of J with respect to each of the analysed values is zero. Computing the gradient of (11.3) we get

$$\nabla_X J = 2B^{-1}(X - X_B) + 2R^{-1}(Y - HX).$$

Setting this to zero we can deduce an expression for X formally identical to (11.1), with the gain matrix again given by (11.2). However, 3D-Var solves the minimisation problem directly, avoiding computation of the gain matrix. The minimum of the cost function is found using a descent algorithm such as the conjugate gradient method. Lorenc (1986) showed that the solution is formally identical to that obtained using OI. However, since the approximations made in implementing the two methods are different, the resulting analyses may also differ. Variational analysis is today becoming the most popular method of data assimilation, as it is the method most suited for assimilation of satellite data.

The analysed fields produced by OI are normally out of balance and an initialisation, such as NNMI or DFI, must be applied to rectify this defect. In variational assimilation, it is straightforward to add to the cost function a 'penalty term' to constrain the analysis. When the global linear balance equation was added as a weak constraint to the National Centers for Environmental Protection (NCEP) analysis, it became unnecessary to perform a separate initialisation during the assimilation cycle (Parrish and Derber, 1992).

The 3D-Var method has enabled the direct assimilation of satellite radiance measurements. The error-prone inversion process, whereby temperatures are deduced from the radiances before assimilation, is thus eliminated. Quality control of these data is also easier and more reliable. As a consequence, the accuracy of forecasts has improved markedly since the introduction of variational assimilation. For several decades, forecasts for the Southern Hemisphere were very poor in comparison to those for the Northern Hemisphere. The skill of medium-range forecasts is now about equal for the two hemispheres. This is due to better satellite data assimilation. Satellite data are essential for the Southern Hemisphere as conventional data are in such short supply.

Inclusion of the time dimension

Whereas conventional meteorological observations are made at the main synoptic hours, satellite data are distributed continuously in time. To assimilate these data, it is necessary to perform the analysis over a time interval rather than for a single moment. Four-dimensional variational assimilation, or 4D-Var for short, uses all the observations within an interval $t_0 \leq t \leq t_N$. The cost function has a term J_B measuring the distance to the background field X_B at the initial time t_0, just as in 3D-Var. It also contains a summation of terms measuring the distance to observations at each time-step t_n in the interval $[t_0, t_N]$:

$$J = J_B + \sum_{n=0}^{N} J_O(t_n)$$

where J_B is defined as for 3D-Var and $J_O(t_n)$ is given by

$$J_O(t_n) = (\mathbf{Y}_n - \mathbf{H}_n \mathbf{X}_n)^T \mathbf{R}_n^{-1} (\mathbf{Y}_n - \mathbf{H}_n \mathbf{X}_n) \tag{11.4}$$

The state vector $\mathbf{X}_n$ at time t_n is generated by integration of the forecast model from time t_0 to t_n, written $\mathbf{X}_n = \mathcal{M}_n(\mathbf{X}_0)$. The vector $\mathbf{Y}_n$ contains the observations valid

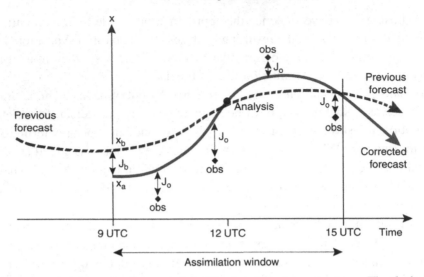

Figure 11.4 Schematic diagram of 4D-Var for a single parameter x. The dashed curve is the evolution for a forecast from an earlier time. The solid curve is from the new analysis $\mathbf{X}_a$. This analysis is chosen to minimise the discrepancies between the model evolution and the observations within the assimilation time window (here a six-hour period). (© ECMWF)

at time t_n. Just as the observation operator had to be linearised to obtain a quadratic cost function, we linearise the model operator $\mathcal{M}_n$ about the trajectory from the background field, obtaining what is called the *tangent linear model* operator $\mathbf{M}_n$. Then we see that 4D-Var is formally similar to 3D-Var with the observation operator $\mathbf{H}$ replaced by $\mathbf{H}_n\mathbf{M}_n$. Just as the minimisation of J in 3D-Var involved the transpose of $\mathbf{H}$, the minimisation in 4D-Var involves the transpose of $\mathbf{H}_n\mathbf{M}_n$, which is $\mathbf{M}_n^\mathrm{T}\mathbf{H}_n^\mathrm{T}$. The operator $\mathbf{M}_n^\mathrm{T}$, the transpose of the tangent linear model, is called the *adjoint model*. The *control variable* for the minimisation of the cost function is $\mathbf{X}_0$, the model state at time t_0, and the sequence of analyses $\mathbf{X}_n$ satisfies the model equations, that is, the model is used as a *strong constraint* (Sasaki, 1970).

The many technical details involved in a practical implementation of four-dimensional variational assimilation have been omitted in the above sketch. They may be found in numerous scientific papers, for example, Courtier *et al.* (1994). The method is illustrated schematically in Fig. 11.4, which shows how the analysis is chosen to minimise the discrepancies between the model evolution and the observations within the assimilation time window, a six-hour period in the case illustrated. The principal characteristics of 4D-Var are as follows.

- The forecast model is assumed to be free from errors. Clearly, this may cause problems when the forecast is inaccurate.
- Derivation of a tangent linear model and implementation of the adjoint operator $\mathbf{M}^\mathrm{T}$ are required. This is a formidable task for complex models.

- All the observational data for the time interval $[t_0, t_N]$ are assimilated.
- By construction, the sequence of model states $\mathbf{X}_n$ is completely consistent with the equations of motion.
- With the perfect model assumption and identical input data, the 4D-Var analysis at time t_n is identical to the Kalman filter analysis at that time. In this sense, 4D-Var is an optimal assimilation algorithm.

The 4D-Var method finds initial conditions $\mathbf{X}_0$ such that the forecast best fits the observations within the assimilation interval. This removes an inherent disadvantage of OI and 3D-Var, where all observations within a fixed time window – typically of three hours – are assumed to be valid at the analysis time. The introduction of 4D-Var at the European Centre for Medium-Range Weather Forecasts led to a significant improvement in the quality of operational medium-range forecasts.

11.3 Progress in computing

Although the ENIAC was five orders of magnitude faster that a human computer, it was severely limited in other ways. It had just 20 words of high-speed memory, so that both the data and program instructions had to be repeatedly read in from punched cards. The IAS computer was more powerful, but still very limited. However, advances in computer technology were rapid. Magnetic core memory, introduced in the IBM 704, was a vital step in making operational NWP feasible. In the 1960s, transistors replaced thermionic valves and, some years later, multitudes of transistors fabricated on a single chip or integrated circuit enabled enormous increases in computer power. The development of vector machines and more recently of massively parallel processors has led to further dramatic growth in number-crunching power. The increase in computer power has been encapsulated in an empirical relationship known as Moore's Law. The essence of this is that the growth of processing power is exponential, with a doubling time of about 18 months. We will illustrate this by considering the sequence of computer platforms at the UK Met Office over the past 50 years.

In 1959, the Met Office purchased its first in-house computer, a Ferranti Mercury, nicknamed 'Meteor'. A two-level model covering the north-eastern Atlantic and Europe on a 320 km grid was run out to 36 hours ahead. The model was experimental and was not used for operational forecasting. Each forecast took several hours – Meteor performed just 3000 calculations per second (or Flops: floating-point operations per second). This was insufficiently fast for operational use. In 1965, the Met Office acquired its second computer, an LEO KDF 9 (English Electric). Its speed was about 5×10^4 calculations per second and its memory size about 12 kWords (Hinds, 1981). On this computer a three-level, quasi-geostrophic model (Bushby and Whitelam, 1961) with a horizontal resolution of about 300 km was run twice daily to 30 hours ahead. The first computer forecasts for use in operations were produced in November 1965.

In 1972, an IBM 360/195 was purchased by the Office. Its speed was about four million calculations per second and its main memory 250 kWords allowing the Bushby–Timpson ten-level primitive-equation model to be run. The resolution was again about 300 km, but with the new computer it was possible to run both a model for the Northern Hemisphere and a nested model for the North Atlantic and European area at 100 km resolution. This computer allowed the numerical model to be refined, bringing increased forecast accuracy, and the 48-hour root-mean-square 500 hPa height error was reduced from 76 m in 1972, to 52 m in 1982. A CDC Cyber 205 was purchased that year and a 15-level primitive-equation global model was implemented. The Cyber 205 operated at 2×10^8 calculations per second (200 MFlops) and had a 1 MWord memory. Global forecasts for up to six days ahead were now produced routinely. The resolution of the model was about 150 km, and the resolution of the nested limited-area model about 75 km. An 11-level version of the global model, with a 260 km resolution, was used for climate modelling.

The multiple processor Cray Y-MP C90, purchased in 1991, had a speed of 10^9 calculations per second (1 GFlops) and a 256 MWord memory. A mesoscale model with 17 km resolution was nested within the new 19-level global model, the *Unified Model* (Cullen, 1993). The resolution of the global model was 90 km. Six years later, the first massively parallel processor was installed in the Office. This was a Cray T3E-900, with 696 processors and a peak processing speed of about 3×10^{11} calculations per second (300 GFlops). It was later expanded to 856 processors, and ran the Unified Model on a 65-km global grid with 30 levels. The nested mesoscale grid had 12 km resolution and 39 levels, producing four forecasts per day to 36 hours ahead.

Following relocation to its new headquarters in Exeter in 2004, the Met Office changed computers again, acquiring an NEC SX-6 supercomputer and, the following April, a more powerful NEC SX-8. The SX-6 and SX-8 have fewer, but much more powerful, processors than the T3E. The SX-6 and SX-8 machines are divided into 'nodes' with each node having eight processors. The SX-8 has 16 nodes, and there are two SX-6 clusters; one with 19 nodes, and the other with 15 nodes. The combined system has a sustained power of about 20 TeraFlops (20 trillion calculations per second).

Processing speed of the Met Office computers from 1959 to 2005 is shown in Fig. 11.5; the logarithm of the speed is plotted against the year (some intermediate platform upgrades are omitted). The thin line indicates exponential growth according to Moore's Law, which states that processor speed (actually, component

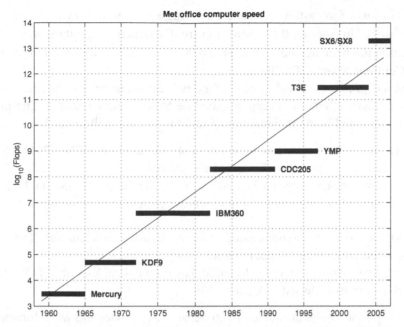

Figure 11.5 Processing speed of Met Office computers from 1959 to 2005 (common logarithm of speed in operations per second). The thin line indicates exponential growth according to Moore's Law. (Data from http://www.metoffice.gov.uk)

density) doubles approximately every 18 months. Clearly, the actual growth in computer power is in general agreement with this empirical law.

In parallel with the advances in computer hardware, increasingly efficient numerical methods of integrating the equations have been developed, and meteorologists have been at the forefront of this work. In addition they have been in the vanguard in devising software for massively parallel processing. As a result of these developments, models of vastly increased complexity and realism are now in common use for numerical weather prediction.

11.4 The European Centre for Medium-Range Weather Forecasts

Perhaps the most important event in European meteorology over the last half-century was the establishment of the European Centre for Medium-Range Weather Forecasts (ECMWF). The mission of 'the Centre' is to deliver weather forecasts of increasingly high quality and scope from a few days to a few seasons ahead. The Centre has been spectacularly successful in fulfilling its mission, and continues to develop forecasts and other products of steadily increasing accuracy and value, maintaining its position as a world leader. The Convention of ECMWF was signed in October 1973 and there were originally 17 Member States (Woods, 2005). The first operational forecasts were made on 1 August 1979. Following profound political changes in Europe, the Centre is currently undergoing enlargement. A new Convention has been agreed and is in the process of ratification. The annual budget of the Centre is about £30 million.

ECMWF produces a wide range of global atmospheric and marine forecasts and disseminates them on a regular schedule to its Member States. The primary products are listed here (explanations of technical terms will follow).

- Forecasts for the atmosphere out to ten days ahead, based on a T799 (25 km) 91-level (L91) deterministic model are disseminated twice per day.
- Forecasts from the Ensemble Prediction System (EPS) using a T399 (50 km) L62 version of the model and an ensemble of 51 members are computed and disseminated twice per day.
- Forecasts out to one month ahead, based on ensembles using a resolution of T159 (125 km) and 62 levels are distributed once per week.
- Seasonal forecasts out to six months ahead, based on ensembles with a T159 (125 km) L40 model are disseminated once per month.

The atmospheric model is coupled to an ocean wave model and wave forecasts are part of the regular dissemination. For a comprehensive discussion of ocean wave modelling, see Janssen (2004). In addition, special forecasts are made to generate boundary values for the limited-area models run in the Member States.

The basis of the NWP operations at ECMWF is the integrated forecast system (IFS). The IFS uses a *spectral representation* of the meteorological fields. Each field is expanded in series of spherical harmonics; for example,

$$u(\lambda, \phi, t) = \sum_{n=0}^{\infty} \sum_{m=-n}^{n} U_n^m(t) Y_n^m(\lambda, \phi)$$

where the coefficients $U_n^m(t)$ depend only on time, and the spherical harmonics $Y_n^m(\lambda, \phi)$ are as introduced in Chapter 3 (p. 51). The coefficients U_n^m of the harmonics provide an alternative to specifying the field values $u(\lambda, \phi)$ in the spatial domain. It is straightforward to transform back and forth between physical space and spectral space. When the model equations are transformed to spectral space, they become a set of equations for the spectral coefficients U_n^m. These are used to advance the coefficients in time, after which the new physical fields may be computed. A continuous field in space requires an infinite series expansion. Just as the continuous field must be replaced by values on a discrete grid to render the problem computationally tractable, the series expansion must be truncated at some point. In the IFS model, the expansion is truncated at a fixed total wavenumber N:

$$u(\lambda_i, \phi_j, t) = \sum_{n=0}^{N} \sum_{m=-n}^{n} U_n^m(t) Y_n^m(\lambda_i, \phi_j).$$

Table 11.1 *Upgrade to the ECMWF integrated forecast system in Spring 2006 (IFS cycle 29r3).*

The spectral resolution is indicated by the triangular truncation number, and the Gaussian grid by the number of points between equator and pole. The number of model levels in the vertical is also given.

	Deterministic model		Ensemble prediction system (EPS)		Monthly forecast (MOFC)	
	Previous	Upgrade	Previous	Upgrade	Previous	Upgrade
Spectral truncation	T511	**T799**	T255	**T399**	T159	**T159**
Gaussian grid	N256	**N400**	N128	**N200**	N80	**N80**
Model levels	L60	**L91**	L40	**L62**	L40	**L62**

This is called *triangular truncation*, and the value of N indicates the resolution of the model. For example, if $N = 512$, the resolution is denoted $T512$. There is a computational grid, called the Gaussian grid, corresponding to the spectral truncation. Since truncation at wavenumber N implies a maximum of N wavelengths around the globe, and since at least two points per wavelength are required, the resolution of the equivalent Gaussian grid is given by the circumference of the Earth divided by twice the truncation N, that is, $\Delta = (2\pi a)/2N$. Since $2\pi a = 4 \times 10^7$ m, we get the simple rule

$$\Delta = \left(\frac{20\,000}{N}\right) \text{ km.} \qquad (11.5)$$

The IFS system underwent a major upgrade in Spring 2006. The horizontal and vertical resolution of its deterministic, ensemble prediction (EPS) and monthly forecasting systems were substantially increased. Table 11.1 compares the spatial resolution of the new model cycle with the previous one. The truncation of the deterministic model is now $T799$, which is equivalent to a spatial resolution of 25 km (it was previously 40 km). The Gaussian grid resolution is conveniently specified by giving the number of points between equator and pole. For the upgraded model, there are 400 such points. The number of model levels in the vertical has been increased by 50 per cent, from 60 to 91. As indicated in the table, the EPS system runs with a horizontal resolution half that of the deterministic model. This reduction in resolution is necessary to enable an ensemble of some 51 forecasts to be completed in reasonable time. The ocean wave forecasting grid was also refined in Spring 2006.

The new Gaussian grid for IFS has about 8×10^5 points. With 91 levels and five primary prognostic variables at each point, about 3×10^8 numbers are required to specify the atmospheric state at a given time. That is, the model has about 300 million degrees of freedom. The computational task of computing forecasts with such high resolution is truly formidable. The Centre carries out its operational programme using a powerful and complex computer system. At the heart of this system is an IBM high performance computing facility (HPCF). Phase 3 of HPCF comprises two identical p690+ clusters. Each cluster consists of 68 computer servers, each having 32 CPUs with a clock frequency of 1.9 GHz. The peak performance is 16.5 TeraFlops for each cluster, so the complete system has a peak performance of 33 TeraFlops or 33 trillion calculations per second.

The ERA-40 re-analysis project

The fruits of an exciting project, the ERA-40 re-analysis, are now available to the scientific community for use in studies of the general circulation, climate change, atmospheric predictability and many other applications. The analyses were produced by ECMWF, working in partnership with a number of institutions. The objective of ERA-40 was to create high-quality global analyses of atmosphere, land and ocean-wave conditions for the past four decades or more using an up-to-date data assimilation system. A comprehensive account of ERA-40, including extensive references to further documentation, is given by Uppala *et al.* (2005).

The production of analyses for the 45-year period from September 1957 to August 2002 was completed in April 2003. The ERA-40 atmospheric analysis was generated using three-dimensional variational data assimilation with six-hourly cycling. The assimilating atmospheric model had T159 spectral truncation in the horizontal (corresponding to 125 km grid spacing) and a 60-level resolution in the vertical, with variables represented up to a pressure of 0.1 hPa. The version of the integrated forecast system that was operational at ECMWF from June 2001 to January 2002 was used, though with some modifications specifically developed for the configuration of the system used for ERA-40. The directly analysed variables were horizontal wind components, temperature, specific humidity and ozone at the 60 model levels, and surface pressure. Many other derived parameters are available; the reanalysis provides us with an unprecedented view of the global circulation of the atmosphere. An atlas of the ERA-40 products has been published (Kållberg *et al.*, 2005).

The re-analysis has enabled sophisticated evaluations of the global observing system to be undertaken. Bengtsson *et al.* (2005) explored the impact of selected observing systems on forecast skill using the system. Analyses were produced for an observing system with only surface-based data (typical of the period prior to 1950), a terrestrial-based observing system with surface and radiosonde data (typical of the

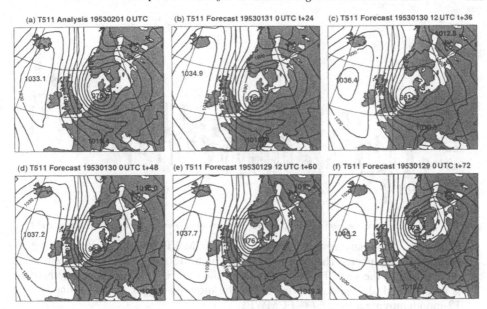

Figure 11.6 The Dutch Storm. (a) Analysis of mean sea-level pressure at 00 UTC on 1 February 1953. Forecasts of various ranges, verifying at the same time: (b) 24-hour, (c) 36-hour, (d) 48-hour, (e) 60-hour, (f) 72-hour forecasts. Analysis and forecasts based on IFS, cycle 25r4. (From Jung *et al.* (2003), © ECMWF)

period 1950–1979), and a satellite-based observing system consisting of surface pressure and satellite observations. Using the 500-hPa geopotential height as a representative field, the terrestrial system in the Northern Hemisphere extratropics was only slightly inferior to the control system (which used *all* observations for the analysis). In the Southern Hemisphere, the forecast skill was dominated by the satellite data, with overall skill comparable to that of the Northern Hemisphere. For the tropics, the information content of the terrestrial and satellite systems was about equal and complementary. Bengtsson *et al.* also studied predictability by comparing how forecasts using different observing systems deviated from each other over time. Their results indicated a potential for a further increase in predictive skill of one to two days in the extratropics of both hemispheres, and a potential for a major improvement of many days in the tropics.

The ERA-40 system has made possible the detailed study of historical storms. Three major European storms of the twentieth century were re-forecast by Jung *et al.* (2003) using the system. One of these, the 'Dutch Storm', was the storm that hit East Anglia and the Netherlands on the night of 31 January 1953. Onshore winds averaged 50–60 knots for a six to nine-hour period prior to high tide, causing the sea level to rise to heights not experienced for many centuries. The devastating floods that followed caused great loss of life. In Fig. 11.6 (panel *a*) the sea-level pressure analysis for 00 UTC on 1 February 1953 is shown. This computer analysis

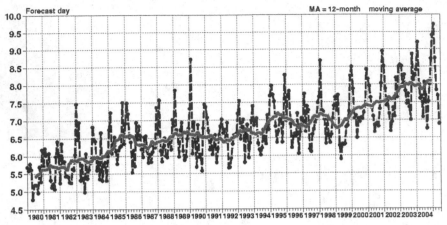

Figure 11.7 Anomaly correlation (AC) of the ECMWF 500 hPa forecast for the Northern Hemisphere. The time in days at which the AC falls to 60 per cent is plotted against the year. This is a measure of the number of days for which there is predictive skill. The dashed line shows the monthly values, the solid line is a 12-month moving average. (© ECMWF)

agrees well with hand analyses produced in real time. A deep depression (976 hPa) is centred over the German Bight, with a strong pressure gradient over the North Sea, corresponding to the onshore gales in the Netherlands. The remaining panels of Fig. 11.6 show forecasts, all verifying at 00 UTC on 1 February 1953, of range 24 hours (panel *b*), 36 hours (panel *c*), 48 hours (panel *d*), 60 hours (panel *e*) and 72 hours (panel *f*). All forecasts have an intense storm in the region and, up to 60 hours, its position is well predicted. Thus, the current ECMWF forecasting system with only the observations available in 1953 is capable of giving warning of this catastrophic event several days in advance. It may be idle to speculate that many lives would have been saved were such a system available at that time. But it is beyond doubt that the forecasts generated by modern NWP systems do save many lives through enabling early precautions when extreme weather is predicted.

Verification of ECMWF forecasts

Forecast skill has improved dramatically in recent decades. Verification scores for the ECMWF model over a period of more than 20 years are shown in Fig. 11.7. The forecast parameter is the 500 hPa height for the Northern Hemisphere. The agreement between the forecast difference from the climate mean value and the observed difference from climate is expressed as the *anomaly correlation* (AC). The higher this score the better; by general agreement, values in excess of 60 per cent imply skill in the forecast. The time in days at which the AC falls to 60 per cent is plotted against the year. The dashed line shows the monthly values,

Anomaly correlation of 500 hPa height forecasts

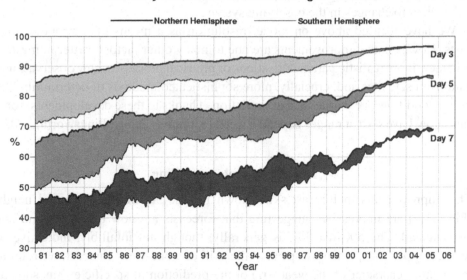

Figure 11.8 Annual running mean anomaly correlation coefficients (%) of 3-, 5- and 7-day 500 hPa ECMWF height forecasts for the extratropical Northern and Southern Hemispheres for the period 1980–2001. The heavy lines are for the Northern Hemisphere and the thin lines for the Southern Hemisphere. The shading shows the difference in scores between the two hemispheres. (Update of Simmons and Hollingsworth (2002), © ECMWF)

the solid line is a twelve-month moving average. The graph shows that in 1980 – the beginning of the period – there was skill in predicting 500 hPa heights out to 5.5 days ahead. Predictive ability has improved steadily over the past 25 years, and there is now skill out to eight days ahead. This record is confirmed by a wealth of other data. Predictive skill has been increasing by about one day per decade, and there are reasons to hope that this trend will continue for several more decades.

In Fig. 11.8, the twelve-month running mean anomaly correlations (in percentages) of the three-, five- and seven-day 500 hPa height forecasts are shown for the extratropical Northern and Southern Hemispheres (Simmons and Hollingsworth, 2002). The heavy lines are for the Northern Hemisphere and the thin lines for the Southern Hemisphere. The shading shows the difference in scores between the two hemispheres. The plots show a continuing improvement in forecast skill, especially for the Southern Hemisphere. By the turn of the millennium, the skill was comparable for the two hemispheres. In the corresponding scores for forecasts from the ERA-40 re-analysis (Uppala *et al.*, 2005) there was hardly any trend. The model used in the re-analysis was unchanged throughout the re-analysis period, whereas it was continually evolving for the operational runs. Thus, we may conclude that the improvements in the operational forecasts since 1980 are primarily due to improvements in the techniques of numerical weather analysis and prediction rather than to changes in the observing system.

We have focused above on model resolution as a means of improving forecasts. However, the improvements are due to many other factors: better numerical schemes, more realistic parameterisations of physical processes, new observational data from satellites and, crucially, more sophisticated methods of determining the initial conditions, i.e. variational assimilation. Thanks to these developments, forecasts now have skill at ranges beyond a week (Simmons and Hollingsworth, 2002).

11.5 Meso-scale modelling

The improvements in forecast skill shown for the 500 hPa geopotential height (Fig. 11.7) are impressive, suggesting that forecasts are now useful beyond one week ahead. The 500 hPa flow is generally, though not infallibly, indicative of weather type. Thus, when we speak of the skill of a one-week forecast, we refer to the general character of the weather, not the prediction of specific events such as thunderstorms or tornadoes. Most people are not interested specifically in the mid-tropospheric flow, but in the actual weather near the Earth's surface. Local weather is often dominated by small-scale processes such as convection, which cannot be explicitly represented in models with grid boxes of size 10 km or more. Richardson alluded to this problem when he considered the effects of eddy motions. The general effect of cumulus convection can be represented by an averaging process, but 'unfortunately this does not help us for example to say whether it will hail or not on Mr X's field' (*WPNP*, p. 65). Progress in forecasting local weather has been much less impressive than synoptic scale prediction. Meso-scale forecasting (say, at scales below 10 km) is much more difficult and may be seen as one of the current grand challenges for meteorology.

The Bushby–Timpson ten-level model was 'one of the first attempts to predict weather, as distinct from pressure patterns and vertical velocity' (Bushby, 1986). It was hoped to simulate frontal rain with this model. However, with a grid-size of 100 km, the success was very limited. A much finer grid is required to properly simulate frontal structures and, to represent cumulus convection explicitly, a grid of just a few kilometres is needed. Even with the powerful computers available today, such resolution cannot be achieved over the entire globe, so *limited area models* are used. Also known as regional models or meso-scale models, these have artificial boundaries where the values of the model variables are provided by a global model run on a coarser grid. This *nesting* procedure is employed in many national meteorological services (NMS) for operational short-range forecast guidance, with the regional model configured to provide high resolution over the geographical area of interest. Nested systems provide local control and autonomy, and have many other attractions:

- The choice of geographical area and model resolution is open.
- There is freedom to run forecasts more frequently.
- Multiple nested systems with several resolutions may be used.
- A more comprehensive range of outputs is available.
- Outputs can be generated at a high time resolution.
- The models may be modified and tuned to local requirements.
- Local modelling expertise is maintained at the NMS.

It is no accident that regional models are ubiquitous. Not only can they provide vital guidance that is beyond the capability of global models, but they produce the high time and space resolution required to drive a large range of other application models essential for serving customer needs. Mesinger (2000) has reviewed the development of limited area modelling, with particular emphasis on the 'Eta-model' used in NCEP.

The treatment of the lateral boundaries of meso-scale models is mathematically complex (there are no real physical boundaries) and is the cause of many practical difficulties. Richardson was aware that a restricted geographical domain would introduce limitations on his algorithm, since the domain would be reduced by a strip one grid interval in width around the boundary for each step forward. 'Only if the table included the whole globe could the repetitions be endless' (*WPNP*, p. 1). To overcome this difficulty, some assumptions had to be made about the behaviour of the atmosphere at the edge of the domain:

The favourite assumption is that the climatological values of the elements . . . are a sufficient representation. This assumption is very easily translated into our numerical process by writing the monthly means around the edge of the table of principal variables. The edge is thus maintained in being, and . . . the wasteful shrinkage of the table at each successive time-step no longer occurs. *(WPNP, p. 153)*

Richardson's proposal to use climatological boundary conditions was employed in some of the earliest NWP models. However, something more accurate is required, and all modern limited area models use values generated by a global model forecast.

The mathematical problem of what values should·be specified on the boundaries was first addressed by Charney *et al.* (1950) and in more detail by Charney (1962). Oliger and Sundström (1978) showed that for any pointwise boundary conditions the (hydrostatic) primitive equations become an 'ill-posed' problem. A pragmatic scheme first proposed by Davies (1976) involved over-specifying the boundary values and relaxing the interior flow towards them over a boundary strip. This method was, and still is, used in the majority of limited area models. McDonald (1997) has provided a comprehensive review of lateral boundary schemes for operational limited area models.

Applications of meso-scale models

The value of the guidance provided by meso-scale models is very high. This is demonstrated by the remarkable range of applications of this information. Amongst the most important application of meso-scale guidance is to provide timely warning of *weather extremes*. Huge financial losses can be caused by gales, floods and other anomalous weather events. Medium-range guidance generally signals large-scale events well in advance. But meso-scale models can give better timing and localisation of extreme events. The higher spatial resolution and use of more recent observational data enables meso-scale models to catch the development of small-scale, localised events that have slipped through the net of medium-range models. The warnings that result from this additional guidance can enable great saving of both life and property.

Transportation, energy consumption, construction, tourism and agriculture are all sensitive to weather conditions. There are expectations from all these sectors of increasing accuracy and detail of short range forecasts, as decisions with heavy financial implications must continually be made. Agriculture is becoming more precise in planning and operations. Meso-scale models provide valuable guidance on ground conditions for work planning, on drying conditions for harvesting, on frost risk for crop damage, on wind conditions for disease dispersal, on rain amounts for a whole range of reasons. Decisions on when to spray crops, when to protect against frost, when to spread fertiliser can be based on objective criteria using short-range forecasts as input.

Winds (or surface stresses) predicted by the meso-scale model are used to drive wave models, which predict sea and swell heights and periods. Forecast charts of the sea state, and other specialised products can be automatically produced and distributed to users. Snowfall is a direct model output, but this can be usefully supplemented by a range of objective indicators of snowfall probability also calculated from meso-scale model output, enhancing the value and usefulness of the NWP guidance.

Limited-area guidance is a crucial component of the input to models for prediction of low-level ozone, pollen levels and other pollution phenomena. Sunburn warnings depend on total ozone levels, and prediction is possible either by treating ozone as a prognostic or passive model variable or, more usually, by using regression techniques to relate stratospheric ozone levels to predictions of atmospheric structure. Trajectories are easily derived from limited area models, either during the course of a forecast or using output fields. These are vital for modelling pollution drift, for nuclear fallout, smoke from forest fires and so on.

Aviation benefits significantly from meso-scale model output. Only the key areas can be mentioned: prediction of hazards (lightning risk, icing risk, clear air turbulence (CAT) and mountain waves), forecast of freezing levels, and of cumulonimbus tops. Very high resolution column models, using input from meso-scale models, are also used to model the boundary layer for applications such as fog prediction at airports. Automatic generation of terminal aerodrome forecasts (TAFs) from meso-scale model and column model outputs enables servicing of a large number of airports from a central forecasting facility.

The most effective short-range forecasts are generated by using NWP products in conjunction with radar output, satellite imagery and other data. Systematic methods of combining a variety of data types are available; SatRep is one such system (see http://www.knmi.nl/satrep or http://www.zamg.at/docu/manual). Model output can be superimposed on satellite imagery enabling assessment of very short range forecasts and thereby giving a measure of forecast quality. Satellite imagery has proved of enormous use to forecasters, but is difficult to incorporate into automatic systems. Pattern recognition techniques have been tried, but with indifferent results. However, a very promising technique has recently emerged. Model fields can be used to construct pseudo-satellite images. A pseudo-water-vapour image generated by the HiRLAM model (Tijm, 2004) is shown in Fig. 11.9 (left panel). The actual Meteosat image for the same time is shown in the right panel. The pronounced circulation north-west of Ireland and the frontal system over France and Germany are evident on both images. Discrepancies between pseudo-satellite images and real images can be used to adjust the automatic analysis, using potential vorticity as an intermediary (Santurette and Georgiev, 2005).

Limited area model guidance has consistently improved in both accuracy and detail over recent years. This guidance is now vital not only for direct use by forecasters, but as driving data for a large range of application models. However, the prediction of exceptional events, such as extreme winds or precipitation, still poses difficulties, and occasionally forecasters fail to provide adequate warning of such events. The main causes of forecast failure are inaccuracies in the initial conditions and imprecision in the model formulation. To address these problems, meteorologists have been moving from deterministic prediction to probabalistic forecasting, and this is discussed in the final section of this chapter.

11.6 Chaos, predictability and ensemble forecasting

In his *Compendium* paper, Charney (1951) indicated that, while forecast skill would improve with better models and faster computers, there would be a limit to predictive skill due to model errors and inaccuracies in the initial conditions. The unpredictability was first considered in detail by Thompson (1957) and elucidated in a simple context by Lorenz (1963). Lorenz discovered that, even for perfect models and almost perfect initial data, the atmosphere has a finite limit of predictability, and he estimated this to be about two weeks. There was an element of serendipity

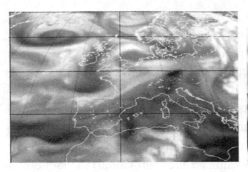

Figure 11.9 Left: model water vapour plus cloud correction image for 06 UTC, 28 August 2004, generated by the HiRLAM model (Tijm (2004), © KNMI). Right: Meteosat water vapour image for the same time (© Eumetsat).

in Lorenz's discovery: he unintentionally performed what we now call an 'identical twin' experiment, when he re-started an earlier computation using rounded intermediate values. He found that the re-run remained close to the original solution for some time but gradually diverged until the two integrations were as different as two randomly chosen solutions. Systems having solutions that depend sensitively on the initial conditions are called *chaotic systems*. Although a full appreciation of chaotic behaviour had to await the advent of computers, this phenomenon was anticipated in a truly visionary way by the French mathematician Poincaré.

Henri Poincaré (1854–1912) was described by the historian of mathematics, E. T. Bell, as the 'Last Universalist', competent in all branches of mathematics. He made major contributions to analysis, algebra, topology, astronomy and theoretical physics. In his study of the three-body problem of celestial mechanics, Poincaré came to realise that even if a physical problem is soluble in principle, practical prediction may be impossible. His understanding of the problem of sensitive dependence on initial conditions, is shown in an essay published in his *Science and Method* (1908):

If we knew exactly the laws of nature and the situation of the universe at the initial moment, we could predict exactly the situation of that same universe at a succeeding moment. But ... it may happen that small differences in the initial conditions produce very great ones in the final phenomena. A small error in the former will produce an enormous error in the latter. Prediction becomes impossible ... .

(quoted from Peterson (1993), who dubbed Poincaré the 'Prophet of Chaos'). Poincaré continued with another illustration of the problem, particularly germane to us:

Why have meteorologists such difficulty in predicting the weather with any certainty? Why is it that showers and even storms seem to come by chance, so that many people think it quite natural to pray for rain or fine weather; though they would consider it ridiculous to ask for an eclipse by prayer?

He understood that developing disturbances occur in regions where the atmosphere is unstable, and that the observations are neither sufficiently comprehensive nor sufficiently accurate to prescribe them completely. Moreover, what appear as very trifling initial perturbations may have considerable effects, leading on occasions to 'terrible disasters'.

Lorenz (1963) introduced a system of three equations that has been studied intensively and that provides a powerful illustration of the phenomenon of chaos. His equations may be written

$$\dot{x} = -\sigma x + \sigma y$$
$$\dot{y} = rx - y - xz \qquad (11.6)$$
$$\dot{z} = -bz + xy.$$

The system describes the time evolution of the three variables x, y and z and the solution may be represented by the trajectory of the point (x, y, z) in a three-dimensional phase space. The parameters may be varied but normally have the values $\sigma = 10$, $r = 28$ and $b = \frac{8}{3}$. The system is dissipative, since the divergence is negative:

$$D \equiv \left(\frac{\partial \dot{x}}{\partial x} + \frac{\partial \dot{y}}{\partial y} + \frac{\partial \dot{z}}{\partial z} \right) = -(1 + \sigma + b) = -13\frac{2}{3} < 0.$$

A volume in phase space with initial value V_0 will decrease exponentially, $V(t) = V_0 \exp(-Dt)$. The system has a bounded globally attracting set of dimension smaller than three, the dimension of the phase space, and all trajectories rapidly approach this attractor. There are equilibria of the system (11.6) at the origin (a hyperbolic saddle-point) and at the *unstable spiral points* $C_1 = (\sqrt{b(r-1)}, \sqrt{b(r-1)}, r-1)$ and $C_2 = (-\sqrt{b(r-1)}, -\sqrt{b(r-1)}, r-1)$. Irrespective of the initial conditions, the solution rapidly settles down to an unending sequence of orbits about the spiral points. It spins in amplifying oscillations about one point for some time, then switches to oscillations about the other. It continues to alternate between these two modes of behaviour, with the number of circuits around each point varying in an erratic manner. The projection of a typical trajectory on the x–z-plane is illustrated in Fig. 11.10. The familiar butterfly-pattern is especially evident in the figure.

We may think of the two lobes of the attractor as representing two distinct weather regimes; for example, westerly zonal flow and blocked flow in middle latitudes. Transitions between the left and right lobes then correspond to transitions between one weather type and another. The irregularity of the transitions between the cycles about the two spiral points means that long-term prediction of the solution

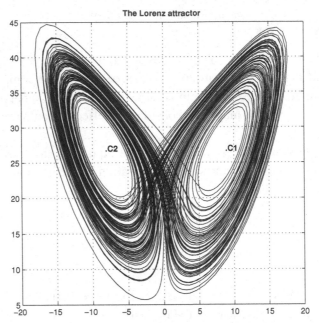

Figure 11.10 The strange attractor of the Lorenz System (11.6) with $\sigma = 10$, $r = 28$ and $b = \frac{8}{3}$. The projection of the trajectory on the x–z-plane is shown. The orbit circulates about the unstable spiral points C_1 and C_2, switching erratically between them.

is impossible. Trajectories starting from almost identical initial states follow similar paths for a short time, but soon diverge. An *identical twin* experiment is shown in Fig. 11.11: the two 'forecasts', started from almost identical initial conditions, soon part company and evolve in completely different ways. Since small errors in the initial conditions are unavoidable, detailed prediction of the long-term evolution is impossible; the system is chaotic.[2]

The trajectory in Fig. 11.10 appears to cross itself. However, this is an illusion: the illustration is a two-dimensional projection of a three-dimensional orbit that never intersects itself. In fact, the orbit spans an intricate set of points with a complex geometric structure. Standard methods of measuring the dimension of this set yield values that are not whole numbers; the set, called a *strange attractor*, has fractional dimension; it is a *fractal*. It was Richardson, in his efforts to measure the lengths of borders and coastlines, who discovered the fractal nature of such curves: in effect, the length of a ragged, indented coastline depends on the 'ruler' used to measure it, increasing indefinitely as the size of the ruler is reduced. This work served as an

[2] For a computer-assisted proof that the Lorenz equations exhibit chaotic dynamics, see Mischaikow and Mrozek, 1995).

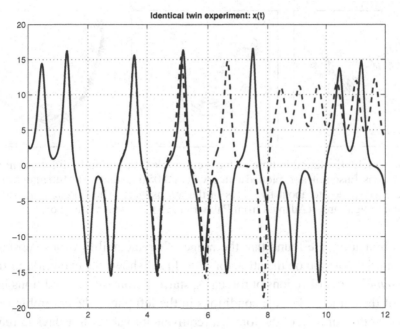

Figure 11.11 Identical twin experiment: evolution of the variable x for two integrations of the Lorenz System (11.6) with almost identical initial conditions.

inspiration to Benoit Mandelbrot (1967, 1982) in his development of the theory of fractals.

Lorenz explained the importance of the attracting set thus:

A strange attractor, when it exists, is truly the heart of a chaotic system. If a concrete system has been in existence for some time, states other than those extremely close to the attractor might as well not exist; they will never occur. For one special complicated system – the global weather – the attractor is simply the climate, that is, the set of weather patterns that have at least some chance of occasionally occurring. *(Lorenz, 1993, p. 50)*

Initialisation is the process of approximating the initial state by a state that is both close to the original state and on or near the attractor. If the original analysis is close to a balanced state, only minor adjustments are necessary. If it is far from balance, larger adjustments will be required. In general, uninitialised analyses are significantly removed from the attractor of the global weather system. In particular, Richardson's analysis was seriously unbalanced and, 'Inevitably, Richardson predicted the violent oscillations that his assumed initial state demanded' (Lorenz, 1993).

Palmer (1993) investigated the physical basis for extended-range atmospheric prediction using the simple Lorenz model as a paradigm for the extra-tropical

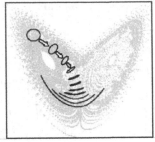

Figure 11.12 Evolution in time of three ensembles starting from different initial conditions. Ensembles are plotted at intervals of 0.05 time units. The full integration time corresponds to about five days in the atmosphere. The Lorenz attractor is shown in light grey. (Adapted from Palmer (1993), © Amer. Met. Soc.)

general circulation. He found that the range of predictability varies substantially with the initial position on the attractor. Fig. 11.12 shows the evolution in time of three *ensembles*, or collections of forecasts, starting from initial conditions in three regions of the attractor. For the conditions in the left panel, the ensemble remains cohesive for the duration of the forecast (equivalent to about five days in real time units). This implies that there is good predictability for initial conditions in this region of the attractor: all the trajectories in the ensemble undergo a transition from one lobe to the other. The ensemble in the centre panel approaches a region where trajectories diverge, and spreads out before the end of the forecast. The initial conditions in the right panel are in this sensitive region, and diverge rapidly: predictive skill is lost almost immediately.

If the atmosphere behaves like the simple three-component model of Lorenz, long-range forecasting must be impossible. But how can we determine whether or not the atmosphere is chaotic? We cannot carry out control experiments: if we disturb the atmosphere in some way, we will never know how it might have behaved in the absence of the disturbance. Our conclusions about the atmosphere are based on identical twin experiments carried out with computer models of the atmospheric circulation.

Almost without exception, the models have indicated that small initial differences will amplify until they are no longer small. There is even good quantitative agreement as to the rate of amplification. ... we are more or less forced to conclude that the atmosphere itself is chaotic.   *(Lorenz, 1993,* p. 102)

Although we cannot claim to have proven that the atmosphere is chaotic, the evidence that it is so is overwhelming:

- Small errors in the synoptic structure of the atmosphere tend to double in about two days.
- Small errors in the finer structure, on the scale of cumulonimbus clouds, tend to grow much more rapidly, doubling within a few hours or less.

- As errors in the fine structure grow, they induce errors at larger scales. Thus, after a day or so there are inevitably errors in the synoptic scales, which will grow just as if they had been present initially.
- Halving the errors of the smallest scales might extend the range of acceptable prediction of the synoptic pattern by only an hour or so.

Of course, this view may be unduly pessimistic: some predictive skill may be retained even at monthly and seasonal timescales when forcing factors, such as sea surface temperature, are both dominant and slowly varying.

Ensemble prediction systems

Until relatively recently, weather forecasts were based on a single deterministic model integration. In view of the chaotic nature of the atmosphere, there are inherent difficulties with this approach: forecasts are sensitive to small initial perturbations and this sensitivity is flow-dependent. So, in some cases analysis errors will grow rapidly whilst in others they will not. It is vital for forecasters to know in advance whether the atmosphere is in a predictable or unpredictable state, so that the reliability of the forecast can be gauged. In practice, forecasters have studied the inconsistencies between forecasts with different analysis times, and have compared forecasts from different centres, to assess the reliability of the numerical guidance. Such so-called 'poor man's ensembles' are much too limited in size to give reliable probability forecasts.

Since the early 1990s, a more systematic method of providing an a-priori measure of forecast skill has been operational at both ECMWF and NCEP in Washington. For a description of the American system, see Kalnay, 2003, §6.5. In the Ensemble Prediction System (EPS) of ECMWF, an ensemble of forecasts (51 in the present system) is performed, each starting from slightly different initial conditions. The sizes of the initial perturbations are comparable to uncertainties in the analysis. A large spread in the ensemble of forecasts indicates low predictability whereas low spread is an indication that predictability is high. The EPS approach is *probabilistic* rather than *deterministic*. We consider the initial state to be specified not precisely, but by a probability density function (PDF) (see Fig. 11.13). The objective is to predict the evolution in time of the PDF. The heavy solid curve in the figure is the deterministic forecast, the heavy dashed curve represents the actual evolution. The thin curves are members of a forecast ensemble, starting from perturbed initial conditions. From them, the PDF at the future time $D + n$ is estimated. This enables the forecaster to provide an objectively determined likelihood of particular meteorological events, such as extreme rainfall. For the case illustrated in Fig. 11.13, the PDF becomes bimodal, suggesting that either of two scenarios is possible. This

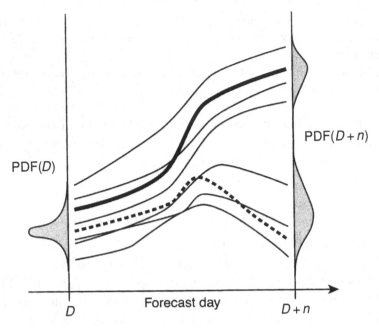

Figure 11.13 Schematic diagram of ensemble prediction. PDF(D) is the probability density function at the initial time, and PDF($D + n$) the corresponding function n days later. Heavy solid curve: deterministic forecast. Heavy dashed curve: actual evolution. Thin curves: ensemble members starting from perturbed initial conditions. (© ECMWF)

might correspond, for example, to a depression passing north or south of the forecast site. The deterministic forecast is in one lobe of the PDF, suggesting one scenario. The actual evolution is towards the other lobe – the deterministic forecast is poor in this case, but the ensemble represents both possibilities well.

To illustrate the value of EPS we consider the prediction of a particularly severe storm that crossed France on 28 December 1999. The mean sea-level pressure (MSLP) analysis for 12 UTC on that day (Fig. 11.14, panel a) shows the storm centred over France (the contour interval is 5 hPa, with shading for values below 980 hPa). The 60-hour T319L60 deterministic forecast verifying at that time is in panel b; it is seriously in error. The EPS T159L40 control forecast (panel c) is similarly poor. The average of all the EPS forecasts (panel d) is no better. However, individual ensemble members, five of which are shown in panels e to i, show a wide variation. The first member (panel e) had a position error of 138 km and an intensity error of only 4 hPa. The operational forecasters in possession of this guidance were able to give good warning of severe weather risk, despite the poor deterministic forecast.

The EPS meteogram, or EPSgram for short, is a very valuable and accessible means of displaying ensemble information for a specific location. It is a *probabilistic*

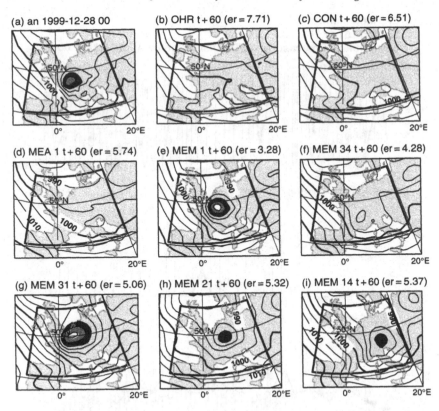

Figure 11.14 Ensemble prediction of the French storm of 28 December 1999. (a) The MSLP analysis at 00 UTC, 28 December 1999. (b) The T319L60 deterministic 60 hour forecast verifying at analysis time. (c) The EPS T159L40 control forecast. (d) The EPS ensemble mean forecast. (e)–(i) Five ensemble members. The contour interval is 5 hPa, with shading for values below 980 hPa. (© ECMWF)

meteogram, indicating the evolution in time of a set of parameters. The ensemble spread, which provides a measure of forecast confidence, is indicated by the range of forecast values. An example is shown in Fig. 11.15. It predicts overcast but mainly dry conditions for the coming week. However, the spikes in the precipitation panel suggest a slight risk of significant precipitation on Sunday 7 August. Probabilistic information of this type is of tremendous value to the forecaster and also for direct use in objective models for a wide range of applications. EPSgrams for a selection of capital cities are available on the ECMWF website (http://www.ecmwf.int) and are updated twice daily.

Weather forecasts have errors because the initial conditions are uncertain and the models are imperfect. Ensembles provide a method of estimating the flow-dependent development of initial uncertainties during the forecast period. Model uncertainties should also be allowed for in the forecast system; these are particularly

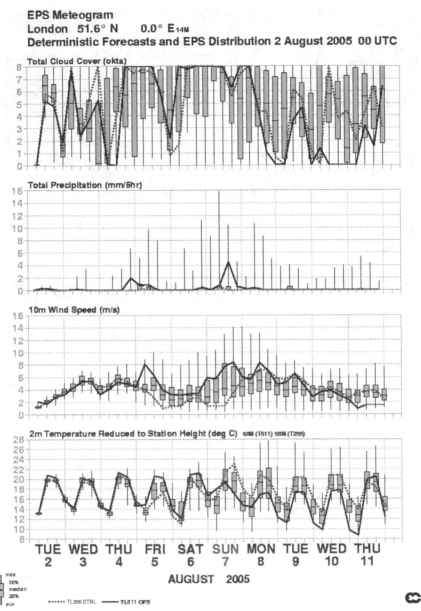

EPS Meteogram
London 51.6° N   0.0° E14M
Deterministic Forecasts and EPS Distribution 2 August 2005 00 UTC

Figure 11.15 EPSgram prediction for London, from analysis time 00 UTC on 2 August 2005. The spikes indicate the full range of the ensemble values, the rectangles the interval around the median of 50 per cent of the ensemble values. The four panels show cloud cover, precipitation, wind speed and temperature. (© ECMWF)

important on monthly to seasonal timescales. The original method was to use multi-model ensembles, running several models with different error characteristics. The multi-model seasonal prediction system at ECMWF comprises ensemble forecasts from the coupled systems of the Met Office, Météo-France and ECMWF. Estimates of the predictability of El Niño with this system suggest some skill at a range of a year ahead.

A problem with conventional ensembles is that they are consistently under-dispersive: whilst initial error growth is realistic, the spread of the ensembles at longer time ranges is not as great as theory would indicate. A relatively new approach is to represent unresolved physical and dynamical processes by stochastic schemes (Palmer *et al.*, 2005). Stochastic parameterisations are currently under investigation in weather forecasting and climate models. They do a better job than conventional bulk parameterisations in simulating the short-wave energy spectrum. Buizza *et al.* (1999) developed and tested a stochastic parameterisation scheme for the ECMWF model. They used the simple stochastic parameterisation:

$$\dot{\mathbf{X}} = \mathbf{D} + (1 + \varepsilon)\mathbf{P}$$

where $\dot{\mathbf{X}}$ denotes the tendency of the model state, $\mathbf{D}$ denotes dynamical core terms (*e.g.* advection and Coriolis terms) and $\mathbf{P}$ denotes the parameterised tendency associated with sub-grid physical processes. The stochastic variable ε is drawn from a uniform distribution in the interval $[-0.5, +0.5]$ and is held constant over a time range of 6 h and a spatial domain of $10° \times 10°$. Buizza *et al.* found that this scheme had a positive impact on medium-range probability forecast skill scores for precipitation, mainly by increasing the ensemble spread for this variable. There is evidence that both model climatology and forecast skill improve as a result of using these schemes; the future will tell.

12

Fulfilment of the dream

Little did Bjerknes know that Richardson would start to bore the tunnel
just a few months later. Still less could he have imagined that express
trains would be driving throught the tunnel within about forty years.

(Ashford, 1985, p. 80)

12.1 Richardson's explanation of his *glaring error*

There is no doubt that Richardson's outlandish forecast results acted as a deterrent
to others who might have been tempted to continue his work. We have seen that
the unrealistic tendencies arose from a disharmony between the fields of mass and
of motion; the causes of the forecast failure were discussed in detail in §7.4. It is
of interest to examine Richardson's understanding of the reasons why his forecast
failed.

At the outset, Richardson stated that the forecast was 'spoilt by errors in the
initial data for winds' (*WPNP*, p. 2) arising from the irregular distribution of pilot
balloon stations and from the sparsity of upper air data. Throughout the book, he
repeatedly referred to errors in the winds as the cause of the forecast failure. He
discussed only the egregious *pressure tendency*, even though all his other tendency
predictions were also unrealistic. Richardson was, of course, aware that 'spurious
convergence' would yield an unrealistic tendency of pressure but, although he was
fully conversant with the Dines' compensation effect, he never mentioned the lack
of vertical compensation in his data as a contributary cause. Similarly strange is his
omission of any mention of the strongly ageostrophic nature of the initial winds,
which resulted in large momentum tendencies. Of course, he knew the consequences
of such imbalance: in his introductory example (*WPNP*, Chapter 2) he observed
that the initial pressure field might be chosen arbitrarily, but that 'if the assumed
pressure gradients be unnaturally steep, the consequent changes will be perplexingly

243

violent'. Moreover, the winds could be chosen completely independently of the pressure 'with a qualification similar to that mentioned above'. For this example he chose geostrophic winds, but he never discussed the disharmony between mass and wind in his data for 20 May 1910.

Richardson's proposed method of rectifying the forecast process was to smooth the initial winds; we have discussed his five smoothing methods in §9.2. However, we have seen that smoothing the initial winds does not guarantee a noise-free evolution. Put another way, smoothing the winds may not get us closer to the slow manifold, or to the climate attractor. Richardson never considered smoothing of the mass field. Sverre Petterssen related the following anecdote apropos the meeting convened by Bjerknes in Bergen in 1921:

Richardson used to draw isobars which, as seen by Bergen-school eyes, seemed somewhat unorthodox. The philosophy of smooth fields was dominant while Richardson's isobars represented rather the opposite extreme. On one occasion an analyst invited Richardson's attention to the absence of smoothness, but Richardson was quite undisturbed and answered, 'It doesn't matter what they look like as long as we know the values at grid points'.

(Quoted from Platzman, 1968)

This certainly indicates a misplaced confidence in the ability of spot values of pressure to represent the synoptic flow without further adjustment. More significantly, there was no mention anywhere in *WPNP* of the need for a *mutual adjustment of the mass and wind fields*. Richardson's second smoothing method was to take a time average of the wind observations over a period of hours. Had he also suggested applying a similar averaging to the mass field, he would have proposed what was, in effect, a digital filtering initialisation technique. However, he did not do that. We are forced to conclude that Richardson's understanding of the causes of his forecast failure was quite incomplete. Moreover, his claim that smoothing of the initial winds would yield a realistic forecast (*WPNP*, p. 217) is seen to be unsustainable.

In Chapter 2 of *WPNP* Richardson made reference to the tidal theory of the atmosphere but he did not appear to appreciate its relevance: 'Much of tidal theory is applicable, but its interest has centred mainly in forced and free oscillations, whereas now we are concerned with unsteady circulations' (*WPNP*, p. 5). Again in Chapter 4 he referred to Lamb's *Hydrodynamics*, specifically to the section dealing with Laplace's theory of the tides on a rotating globe (see Lamb, 1932, Arts. 213–223). But he made no mention of gravity waves in the atmosphere, nor did he appear to recognise their role in causing his forecast failure. Richardson was a master of numerical analysis, and well understood the problems that arise from combining disparate scales. In his paper on the deferred approach to the limit, he introduced a sample function that is everywhere continuous and differentiable to all orders, $f(x) = \sin x + \sin(100x) + \sin(10000x)$:

The analyst finds it pleasant, but to the computer it is an intractable horror. A step h which is large enough to allow satisfactory progress in exploring the variation of $\sin x$ is far too large to reveal the detail of $\sin(10000x)$. Let us call these rapid oscillations, superposed on much slower variations, by the name 'frills'. *(Richardson, 1927)*

The high frequency gravity wave oscillations superimposed on a quasi-balanced flow are precisely the frills that Richardson spoke of but, unfortunately, he was unable to make the connection between his spurious tendencies and the existence of gravity wave oscillations in the atmosphere. One of the reviewers of *WPNP*, F. J. W. Whipple of the Met Office, actually identified the *waves which are propagated with the velocity of sound* as the culprits (see p. 18 above). It is one of the quirks of history that nobody thought it worthwhile looking more deeply into this at the time. We may suppose that there were so many other obstacles to a practical implementation of numerical weather prediction when *WPNP* was published that the scientists of that time did not regard it as a fruitful area of research.

It would appear that Richardson came to realise the need for adjustment of the initial data only after he had completed his forecast. His explanation of the errors of predicted tendency in terms of spurious values of divergence is incomplete, but it is consistent with the analysis of Margules (see §7.5). Had Richardson been aware of Margules' results, he might well have decided not to proceed with the trial forecast, or sought a radically different approach. It is possible that he realised the significance of Margules' results when he read Exner's book but, in that case, it seems inexplicable that he did not refer to Margules, or to the relevant section of Exner, explicitly. He had completed a Homeric numerical forecast and included it in his book, and Margules' results showed that his approach was, from the outset, doomed to failure. Although such a realisation would have been devastating, one cannot doubt that Richardson would have faced it with honesty. Later, Richardson did realise that his original method was unfeasible. Reference was made in §7.5 to a note (undated) in the *Revision File*, where he wrote that the equation of continuity must be eliminated. He further speculated that the vorticity might be a suitable prognostic variable, but we have no evidence that he or any of his contemporaries pursued this line, which later proved so fruitful in the hands of Rossby and Charney.

The theory of atmospheric fronts was undergoing rapid development in Bergen at the time *WPNP* was being finished. Richardson was aware of this development: Vilhelm Bjerknes visited him in Benson in November 1919 and Richardson participated in two scientific conferences convened by Bjerknes in Bergen, in 1920 and 1921. In the preface to *WPNP*, Richardson wrote that '... in the last two years Prof. V. Bjerknes and his collaborators ... have enunciated the view, based on detailed observation, that discontinuities are the vital organs supplying the energy to cyclones'. Shortly after his visit to Benson, Bjerknes wrote to Robert Wenger, his successor in Leipzig, of a conversation with Richardson about the reasons for

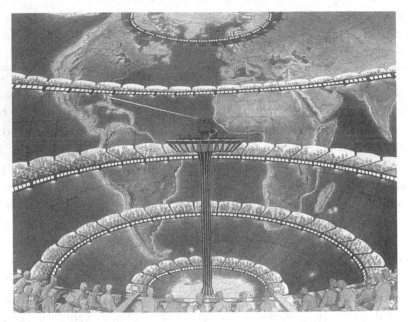

Figure 12.1 An artist's impression of Richardson's forecast factory. (Thanks to artist François Schuiten for permission to reproduce image)

his 'meaningless' forecast results: 'We agreed that the interloper discontinuity was most probably one of the main causes of his failure.' Bjerknes' perspective was perhaps over-influenced by the dramatic progress in frontal theory under way in Bergen: Richardson himself did not make reference, in *WPNP* or elsewhere, to fronts as the cause of his problems. However, he recognised that the numerical process would have to be specially modified to handle such discontinuities. Indeed, he later wrote that, if a second edition of his book were to be produced, he should include a new chapter on the processing of discontinuities.

12.2 The 'forecast factory'

Despite the many obstacles to be overcome before NWP could become a reality, Richardson showed remarkable foresight when he penned his famous *fantasy* of a 'forecast factory' (Fig. 12.1). This has been reproduced widely, but it is so striking that it merits another full exposure:

After so much hard reasoning, may one play with a fantasy? Imagine a large hall like a theatre, except that the circles and galleries go right round through the space usually occupied by the stage. The walls of this chamber are painted to form a map of the globe. The ceiling represents the north polar regions, England is in the gallery, the tropics in the upper circle, Australia on the dress circle and the antarctic in the pit. A myriad computers are at work upon the weather of the part of the map where each sits, but each computer attends only to one equation or part of an equation. The work of each region is coordinated by an official of higher rank. Numerous little "night signs" display the instantaneous values so that neighbouring computers can read them. Each number is thus displayed in three adjacent zones so as to maintain communication to the North and South on the map. From the floor of the pit a tall pillar rises to half the height of the hall. It carries a large pulpit on its top. In this sits the man in charge of the whole theatre; he is surrounded by several assistants and messengers. One of his duties is to maintain a uniform speed of progress in all parts of the globe. In this respect he is like the conductor of an orchestra in which the instruments are slide-rules and calculating machines. But instead of waving a baton he turns a beam of rosy light upon any region that is running ahead of the rest, and a beam of blue light upon those who are behindhand.

Four senior clerks in the central pulpit are collecting the future weather as fast as it is being computed, and despatching it by pneumatic carrier to a quiet room. There it will be coded and telephoned to the radio transmitting station. Messengers carry piles of used computing forms down to a storehouse in the cellar.

In a neighbouring building there is a research department, where they invent improvements. But there is much experimenting on a small scale before any change is made in the complex routine of the computing theatre. In a basement an enthusiast is observing eddies in the liquid lining of a huge spinning bowl, but so far the arithmetic proves the better way. In another building are all the usual financial, correspondence and administrative offices. Outside are playing fields, houses, mountains and lakes, for it was thought that those who compute the weather should breathe of it freely.

Richardson's description is certainly whimsical but it is also remarkably prescient. There are surprising similarities between his forecast factory and a modern massively parallel processor (MPP). Richardson envisaged a large number of processors – 64 000 by his estimate – working in synchrony on different sub-tasks. The fastest computer in the Top 500 list as of June 2005 was the IBM BlueGene/L with 65 536 processors! The silicon-based processing elements of modern computers are incomparably more powerful than the carbon-based 'computers' proposed by Richardson. The IBM machine is rated at 136.8 TFlops (136 trillion calculations per second; see http://www.top500.org). The BlueGene is perhaps nine orders of magnitude faster than Richardson's forecast factory. In the fantasy, the forecasting job is sub-divided, or parallelised, using domain decomposition, a technique often used in MPPs today. Richardson's *night signs* provide nearest-neighbour communication, analogous to message-passing techniques in MPPs. The man in the pulpit, with his blue and rosy beams, acts as a synchronisation and control unit. Thus, while the processing speeds differ by many orders of magnitude, the logical structures of the forecast factory and the MPP have much in common.

The dawn of the atomic era brought with it the need for mathematical computations on a scale greater than ever before. The Manhattan Project had access to the most advanced technology available, though it was primitive by modern standards. The workhorse for scientific computing was an electro-mechanical machine, the Marchand calculator, which could add, subtract, multiply and (with difficulty) divide numbers of up to ten digits. At Los Alamos the computations were organised like a factory assembly line (Gleick, 1992). The staff – mostly the wives of the scientists, working on reduced wages – worked in a large array, like the cogs of a great machine, each computing an individual component of a complex system of equations, cranking the handle of her Marchand and communicating results to her neighbours; the arrangement was analogous to the forecast factory. The output from the production line was a detailed calculation of the behaviour of the expanding ball of fire in a thermo-nuclear explosion. This would hardly have met with the approval of the pacifist Richardson. Some of the early numerical weather forecasts were computed using a man–machine mix or, perhaps more accurately, woman–machine mix, like that in Los Alamos. One such example in Germany, where access to computers was unavailable in the aftermath of the war, was described by Edelmann (see page 201). Fortunately, powerful automatic data processing soon took over the drudgery of such calculations.

12.3 Richardson's dream

We opened with a quotation expressing Richardson's dream: 'Perhaps some day in the dim future it will be possible to advance the computations faster than the weather advances . . . .' The ensuing chapters have described how that dream, utterly fanciful in the dim past, has been fulfilled in a spectacular fashion. Progress in numerical weather prediction has been dramatic and has been of huge benefit to humankind. It has brought us far beyond anything Richardson could have imagined, and continues to develop apace. Satellite systems now observe the atmosphere and oceans continuously, dedicated communication networks distribute weather data at the speed of light, and powerful computer systems using sophisticated numerical algorithms perform prodigious calculations to predict the weather for many days ahead.

   There is a strong symbiosis between numerical weather prediction and theoretical meteorology. Advances in our understanding of the physics and dynamics of the atmosphere and ocean are soon exploited in computer models, and these models themselves provide us with a powerful method of exploring the behaviour of the real atmosphere and ocean. George Cressman, an early pioneer of numerical prediction, once remarked that 'the problems of NWP can be considered to be the problems of all meteorology'. More than ever, this is true today. Numerical weather prediction has now reached a high level of sophistication. Forecasts up to a week or more ahead are of value, and progress is under way in monthly and seasonal prediction.

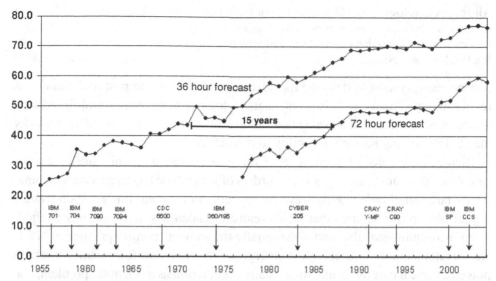

Figure 12.2 Skill of the 36 hour (1955–2004) and 72 hour (1977–2004) 500 hPa forecasts produced at NCEP. Forecast skill is expressed as a percentage of an essentially perfect forecast score. The accuracy of prediction is closely linked to the available computer power; the introduction of new machines is indicated in the figure. Thanks to Bruce Webster of NCEP for the graphic of S_1 scores.

At longer timescales, models of the sort first formulated by Richardson are our best means of anticipating changes in global climate, which may have profound consequences for humanity.

Prior to the computer era, weather forecasting was in the doldrums. Petterssen (2001) described the advances as occuring in 'homeopathic doses'. The remarkable progress in forecasting over the past 50 years is vividly illustrated by the record of skill of the 500 hPa forecasts produced at the National Meteorological Center, now NCEP, as measured by the S_1 score (Teweles and Wobus, 1954). The 36 hour scores are the longest verification series in existence, dating from the very beginning of operational NWP. The skill scores, expressed as percentages of maximum possible skill, have improved steadily over the past 50 years and each introduction of a new prediction model has resulted in further improvement (Fig. 12.2). The sophistication of prediction models is closely linked to the available computer power; the introduction of each new machine is also indicated in the figure. The horizontal bar indicates a 15 year delay for the 72 hour forecast to attain the skill previously attained at 36 hours. This is consistent with the general experience of a one day per-decade increase in forecast skill.

A pioneer of numerical weather prediction, Fred Shuman, concluded his historical review of NWP at the National Meteorological Center thus:

All the meteorological world was watching the work ... [of JNWPU] in the 1950s. Our job was no less than to revolutionize weather forecasting, which had begun almost a century earlier as a centralized operation, and which had not changed much since then in its fundamental processes.

(Shuman, 1989)

It is no exaggeration to describe the advances made over the past half century as revolutionary. Thanks to this work, meteorology is now firmly established as a quantitative science, and its value and validity are demonstrated on a daily basis by the acid test of any science, its ability to predict the future.

Richardson's forecast came to grief through his use of uninitialised data: his calculated pressure tendency was two orders of magnitude too large, due to anomalously large amplitude gravity wave components in his data. Initialisation using a digital filter produces data that yields realistic tendencies. Richardson's methodology was unimpeachable and is essentially the same as current practice in NWP. He was not alone in foreseeing the emergence of numerical forecasting. Bjerknes played a critical part by formulating weather prediction as a scientific problem, and Helmholtz had earlier contributed by completing the system of equations through his developments in thermodynamics. But it was Richardson who actually had the vision and the audacity to put to a practical test what earlier scientists had seen only in a theoretical context. For that alone, he is worthy of our admiration.

Appendix 1

Table of notation

An auxiliary language, regularly structured and easy to learn, would offer an attractive means of communication between people of different ethnic origins. Dozens of artificial languages were constructed between the late nineteenth and mid twentieth centuries, the most noteworthy being Volapük, Occidental, Novial, Interglossa, Interlingua, Ido and Esperanto. Only the last-named attracted a sizeable body of speakers, currently numbered in millions (http://www.uea.org/).

The history of artificial languages is fraught with conflict, controversy and schism. Many believe that the ascendancy of English has removed the need for an auxiliary language; but English is neither regularly structured nor easily learned. Richardson believed that an international language could help in avoiding inter-ethnic wars. Although he chose Ido for his List of Symbols (*WPNP*, p. 223), he also had knowledge of Esperanto (Ashford, 1985, pp. 57, 181).

As Esperanto is the only currently credible candidate to serve as an international auxiliary language, it is chosen for the table below. The principal mathematical symbols used in the text are defined in both English and Esperanto. Entries have been checked against the standard dictionary (PIV, 2005).

Symbol	Meaning	Signifo
$\overline{(\ )}$	Horizontal average	Horizontala mezvaloro
$(\ )_H$	Horizontal component	Horizontala komponanto
$(\ )_{1,...,5}$	Values in vertical layers	Valoroj en vertikalaj tavoloj
$(\ )_S$	Value at Earth's surface	Valora ĉe la Tera surfaco
$(\ )_0$	Constant reference value	Konstanta norma valoro
$(\ )_{geo}$	Geostrophic component	Geostrofa komponanto
A	Analysis error covariance matrix	Kunvarianca matrico de analiza eraroj
B	Background error covariance matrix	Kunvarianca matrico de fonaj eraroj
Br	Balance ratio, $100 \times (N_1/N_2)$	Ekvilibra proporcio, $100 \times (N_1/N_2)$
D	Mass divergence, $\nabla \cdot \rho\mathbf{V}$	Diverĝenco de maso
Γ	Force due to friction	Frota forto
H	Observation operator for analysis	Observa operatoro por analizo

Symbol	Meaning	Signifo
H	Scale height of atmosphere, $\mathfrak{R}T_0/g$	Skala alteco de la atmosfero
J	Analysis cost function	Analiza kostfunkcio
K	Analysis gain matrix	Analiza gajnmatrico
N	Brunt–Väisälä frequency	Brunt–Väisälä frekvenco
N_1, N_2	Noise parameters	Parametroj de bruo
P	Pressure of layer, $\int p\, dz$	Premo de tavolo
R	Observation error covariance matrix	Kunvarianca matrico de observeraroj
$\mathfrak{R}$	Gas constant for dry air	Gaskonstanto de seka aero
R	Density of layer, $\int \rho\, dz$	Denso de tavolo
S	Specific entropy, $c_p \log \theta$	Specifa entropio
T	Temperature	Temperaturo
U	Three-dimensional velocity	Tridimensia rapido
U	Zonal layer momentum, $\int \rho u\, dz$	Zona movokvanto de tavolo
V	Horizontal velocity (horizontal momentum in Ch. 3 and Ch. 4)	Horizontala rapido (horizontala movokvanto en Ĉap. 3 kaj Ĉap. 4)
V	Meridional layer mtum., $\int \rho v\, dz$	Meridiana movokvanto de tavolo
W	Water content of layer, $\int q\, dz$	Akvokvanto de tavolo
X	Solution vector in phase space	Solvovektoro en fazspaco
Y	Projection of **X** on slow sub-space	Projekcio de **X** sur malrapideca subspaco
Z	Projection of **X** on fast sub-space	Projekcio de **X** sur rapideca subspaco
$\mathcal{H}$	Scale factor for height	Skalfaktoro de alteco
$\mathcal{L}$	Scale factor for length	Skalfaktoro de longeco
$\mathcal{P}'$	Scale factor for pressure variation	Skalfaktoro de premvariado
$\mathcal{R}'$	Scale factor for density variation	Skalfaktoro de densvariado
$\mathcal{S}$	Slow manifold	Malrapideca sternaĵo
$\mathcal{T}$	Scale factor for time	Skalfaktoro de tempo
$\mathcal{V}$	Scale factor for horizontal velocity	Skalfaktoro de horizontala rapido
$\mathcal{W}$	Scale factor for vertical velocity	Skalfaktoro de vertikala rapido
$\mathcal{X}$	Phase space of solutions	Fazspaco de solvoj
$\mathcal{Y}$	Slow linear subspace of $\mathcal{X}$	Malrapideca linia subspaco de $\mathcal{X}$
$\mathcal{Z}$	Fast linear subspace of $\mathcal{X}$	Rapideca linia subspaco de $\mathcal{X}$
a	Radius of Earth	Radiuso de la Tero
c	Phase speed of wave	Fazrapido de ondo
c_p	Specific heat at constant pressure	Specifa varmo je konstanta premo
c_v	Specific heat at constant volume	Specifa varmo je konstanta volumeno
f	Coriolis parameter, $\sin \phi$	Parametro de Coriolis, $\sin \phi$
g*	Acceleration due to gravity	Gravita akcelo
g	Apparent acceleration due to gravity	Ŝajna gravita akcelo
g	Magnitude of **g**	Grandeco de **g**
h	Equivalent depth of atmosphere	Ekvivalenta profundo de atmosfero
k	Unit vector in vertical direction	Vertikala unuo-vektoro

Symbol	Meaning	Signifo
k	Wave number in x-direction	Ondonumero en x-direkto
l	Wave number in y-direction	Ondonumero en y-direkto
m, n	Wave numbers (integral)	Ondonumeroj (integraj)
p	Atmospheric pressure	Atmosfera premo
p_0	Reference pressure, 1000 hPa	Norma premo, 1000 hPa
q	Specific humidity	Specifa humido
$\mathbf{r}$	Radius vector from Earth's centre	Radiusvektoro de la tercentro
r	Distance from Earth's centre	Distanco de la tercentro
t	Time	Tempo
u	Eastward component of velocity	Rapidkomponanto orienten
v	Northward component of velocity	Rapidkomponanto norden
w	Vertical component of velocity	Rapidkomponanto supren
x	Distance eastward	Distanco orienten
y	Distance northward	Distanco norden
z	Distance upward	Distanco supren
Φ	Geopotential height	Geopotenciala alto
Ω	Angular velocity of Earth	Angulrapido de la Tero
Ω	Magnitude of Ω	Grandeco de Ω
β	Beta parameter, $2\Omega \cos\phi/a$	Beta-parametro, $2\Omega \cos\phi/a$
γ	Ratio of specific heats, c_p/c_v	Proporcio de specifaj varmoj, c_p/c_v
δ	Velocity divergence, $\nabla \cdot \mathbf{V}$	Diverĝenco de rapido, $\nabla \cdot \mathbf{V}$
ϵ	Lamb's parameter, $(2\Omega a)^2/gh$	Parametro de Lamb, $(2\Omega a)^2/gh$
ζ	Vertical component of vorticity	Vertikala komponanto de vorticeco
θ	Potential temperature	Potenciala temperaturo
κ	Thermodynamic ratio, $\Re/c_p$	Termodinamika proporcio, $\Re/c_p$
λ	Longitude	Longitudo
μ	Sine of latitude, $\sin\phi$	Sinuso de latitudo, $\sin\phi$
ν	Frequency of wave	Ondofrekvenco
ρ	Density	Denso
σ	Non-dimensional frequency	Sendimensia frekvenco
τ	Wave period	Ondoperiodo
ϕ	Latitude	Latitudo
χ	Velocity potential	Rapido-potencialo
ψ	Stream function	Flufunkcio
ω	Vertical pressure velocity, dp/dt	Vertikala premrapido, dp/dt

Appendix 2

Milestones in Richardson's life and career

The main events of Richardson's life and career, and his most important publications, are listed in this appendix. The principal sources of information were Ashford (1985), Gold (1954) and Richardson's *Collected Papers* (LFR I, 1993).

1881 Richardson born 11 October 1881 in Newcastle-upon-Tyne, the youngest of seven children of David Richardson and Catherine Fry.

1894–1899 Pupil at Bootham School, York, where his interest in science was encouraged by excellent teachers.

1898 Passed Matriculation examination for London University.

1898–1890 Attended the Durham College of Science in Newcastle.

1900–1903 Student at King's College, Cambridge.

1903 First Class Honours in Natural Science Tripos, Part I.

1903–1904 Student Assistant in Metallurgy Division, National Physical Laboratory, Teddington.

1904–1905 Junior Demonstrator in Physics, University College, Aberystwyth.

1906–1907 Scientist with National Peat Industries, Newcastle. Solved the differential equations for drainage of water through peat using a graphical method.

1907 Mathematical Assistant to the statistician Karl Pearson at University College London. Devised numerical method for calculating the stresses in masonry dams.

1908–1909 Scientist in the Metrology Division, National Physical Laboratory.

1909 Married Dorothy Garnett, daughter of William and Rebecca Garnett.

1909–1912 Head of Chemical and Physical Laboratory, Sunbeam Lamp Co., Gateshead.

1910 'The approximate arithmetical solution by finite-differences of physical problems involving differential equations, with an application to the stresses in a masonry dam' published in *Phil. Trans. Roy. Soc.*, A **210**, 307–57 (see LFR I, pp. 121–71).

1912 Richardson inspired by sinking of *Titanic* to devise a SODAR (echo detector) device. Also, an ASDIC-type device for use under water. Obtained two patents for these echo-location devices (see LFR I, pp. 995–1009).

1912–1913 Lecturer and Demonstrator, Municipal School of Technology (later UMIST), Manchester.

1913 Richardson joined the Met Office.

1913–1916 Superintendent of Eskdalemuir Observatory. A few publications on geomagnetism.

1916 Completed first draft of his book on NWP, then called *Weather Prediction by Arithmetical Finite Differences*. The draft was communicated to the Royal Society.

1916 Resigned from the Met Office to join the Friends Ambulance Unit (FAU).

1916–1919 Worked as a driver with FAU in the *Section Sanitaire Anglaise* (SSA13) attached to the French Army in the region of Châlons-en-Champagne. The Section came under heavy shelling during the Battle of Champagne in April 1917.

1918 Constructed a working model of the atmosphere, consisting of a basin on a gramophone turntable. Essay, 'The mathematical psychology of war', printed and distributed privately.

1919 Became a Fellow of the Royal Meteorological Society. Re-joined the Met Office.

1919–1920 Worked on numerical forecasting at Benson Observatory, where W. H. Dines was in charge. Revised his book on NWP. Developed some exotic observing methods and instruments: cracker balloons, lizard balloons, shooting spheres.

1920 'The supply of energy from and to atmospheric eddies' published in *Proc. Roy. Soc.*, A **97**, 354–73 (see LFR I, pp. 278–97).

1920 Resigned once more from the Met Office, when it came under the Air Ministry.

1920–1927 Lewis and Dorothy Richardson adopt three children, Olaf (b. 1916), Stephen (b. 1920) and Elaine (b. 1927).

1920–1927 Lecturer in Physics and Mathematics at Westminster Training College, London.

1922 *Weather Prediction by Numerical Process* published by Cambridge University Press.

1925 Qualified as B.Sc. (pass) in Psychology (with mathematics). Awarded a Second Class Honours B.Sc. in 1929.

1926 'Atmospheric diffusion shown on a distance-neighbour graph' published in *Proc. Roy. Soc., London*, A **110**, 709–37 (see LFR I, pp. 523–51).

1925 Participated in conference of the International Commission for Investigation of the Upper Air (ICIUA) in London.

1926 Elected a Fellow of the Royal Society. Awarded a D.Sc. by London University for his published work.

1926 Richardson makes a deliberate break with meteorology, to study quantitative psychology (mathematical psychology).

1927 Participated in conference of ICIUA in Leipzig and in the Assembly of the International Association for Meteorology and Atmospheric Physics in Prague.

1927 'The deferred approach to the limit' published in *Phil. Trans. Roy. Soc.*, A **226**, 299–349 (see LFR I, pp. 627–77).

1929–1940 Principal of Paisley Technical College.

1936 Richardson turns to intensive peace studies.

1936 Godske visits Richardson in Paisley.

1939 Monograph *Generalized Foreign Politics* published. Richardson makes fact-finding trip to Danzig as war looms.

1940 Takes early retirement (aged 59) to pursue his peace studies.

1943 Richardson and his wife move to Hillside House, Kilmun, near Holy Loch.

1947 *Arms and Insecurity* published in microfilm form.

1948 Richardson's diffusion (parsnips) experiment with Henry Stommel (see Ashford, 1985, pp. 211–13).

1947 *Statistics of Deadly Quarrels* published in microfilm form.

1953 Richardson died in his sleep, 30 September.

Appendix 3

Laplace tidal equations: separation of variables

In this appendix we investigate solutions analogous to those studied in §3.1 but allowing for a more general vertical structure. Specifically, we do not require the vertical velocity to vanish. Let us assume that the motions under consideration can be described as small perturbations about a state of rest, in which the temperature is a constant, T_0, and the pressure $\bar{p}(z)$ and density $\bar{\rho}(z)$ vary only with height. The basic state variables satisfy the gas law and are in hydrostatic balance:

$$\bar{p} = \Re\bar{\rho}T_0, \qquad \frac{\partial \bar{p}}{\partial z} = -g\bar{\rho}.$$

The variations of mean pressure and density follow immediately:

$$\bar{p}(z) = p_0 \exp(-z/H), \qquad \bar{\rho}(z) = \rho_0 \exp(-z/H),$$

where $H = p_0/g\rho_0 = \Re T_0/g$ is the scale-height of the atmosphere. Let u, v, w, p and ρ denote variations about the basic state, each of these being a small quantity. We assume the Earth's surface is perfectly flat, so that $w = 0$ there. If all the quadratic terms are omitted, the horizontal momentum, continuity and thermodynamic equations given on page 37 become

$$\frac{\partial u}{\partial t} - fv + \frac{\partial}{\partial x}\left(\frac{p}{\bar{\rho}}\right) = 0 \tag{3.1}$$

$$\frac{\partial v}{\partial t} + fu + \frac{\partial}{\partial y}\left(\frac{p}{\bar{\rho}}\right) = 0 \tag{3.2}$$

$$\frac{\partial \rho}{\partial t} + \nabla\cdot\bar{\rho}\mathbf{V} + \frac{\partial \bar{\rho}w}{\partial z} = 0 \tag{3.3}$$

$$\frac{1}{\gamma\bar{p}}\frac{\partial p}{\partial t} - \frac{1}{\bar{\rho}}\frac{\partial \rho}{\partial t} + \frac{\kappa}{H}w = 0 \tag{3.4}$$

The hydrostatic and state equations are used to express the thermodynamic equation as

$$\bar{\rho}\frac{\partial}{\partial t}\left\{\frac{\partial}{\partial z}\left(\frac{p}{\bar{\rho}}\right) - \frac{\kappa}{H}\left(\frac{p}{\bar{\rho}}\right)\right\} + \left(\frac{\kappa g}{H}\right)\bar{\rho}w = 0. \tag{3.5}$$

256

Now $\bar{\rho}w$ can be eliminated between this equation and the continuity equation (3.3) to give, after some manipulation:

$$\frac{\partial}{\partial t}\left[\frac{1}{\bar{\rho}}\frac{\partial}{\partial z}\bar{\rho}\frac{\partial}{\partial z}\left(\frac{p}{\bar{\rho}}\right)\right] - \frac{\kappa g}{H}\nabla\cdot\mathbf{V} = 0. \tag{3.6}$$

Now let us assume that the horizontal and vertical dependencies of the perturbation quantities are separable:

$$\left\{\begin{array}{c} u \\ v \\ p/\bar{\rho} \end{array}\right\} = \left\{\begin{array}{c} U(x,y,t) \\ V(x,y,t) \\ P(x,y,t) \end{array}\right\} Z(z) .$$

Then (3.6) may be written in the form

$$\left[\frac{1}{\bar{\rho}}\frac{\partial}{\partial z}\bar{\rho}\frac{\partial Z}{\partial z}\right]\frac{\partial P}{\partial t} = \left(\frac{\kappa g}{H}\right)Z\nabla\cdot\mathbf{V}$$

But this is immediately convertable to the separable form

$$\left[\frac{1}{\bar{\rho}}\frac{\partial}{\partial z}\bar{\rho}\frac{\partial Z}{\partial z}\right]\bigg/\frac{\kappa g}{H}Z = \frac{\nabla\cdot\mathbf{V}}{\partial P/\partial t}. \tag{3.7}$$

The left side of (3.7) is a function only of z while the right side depends only on x, y and t; thus, both must be constant. This constant has the dimension of inverse square velocity and we write it as $-1/gh$ where h is a length, called the equivalent depth. The left side gives the vertical structure equation for Z:

$$\left[\frac{1}{\bar{\rho}}\frac{\partial}{\partial z}\bar{\rho}\frac{\partial Z}{\partial z}\right] + \left(\frac{\kappa}{hH}\right)Z = 0. \tag{3.8}$$

The right side, together with the momentum equations, yields the system

$$\frac{\partial U}{\partial t} - fV + \frac{\partial P}{\partial x} = 0 \tag{3.9}$$

$$\frac{\partial V}{\partial t} + fU + \frac{\partial P}{\partial y} = 0 \tag{3.10}$$

$$\frac{\partial P}{\partial t} + gh\nabla\cdot\mathbf{V} = 0 \tag{3.11}$$

This is a set of three equations for the three dependent variables U, V and P. They are mathematically isomorphic to the linear shallow water equations with a mean depth h.

The vertical structure equation

We consider solutions of the vertical structure equation (3.8). To simplify it, we define $\hat{Z} = \exp(-z/2H)Z$ which yields

$$\frac{d^2\hat{Z}}{dz^2} + \left\{\frac{\kappa}{hH} - \frac{1}{4H^2}\right\}\hat{Z} = 0. \tag{3.12}$$

The character of the solution depends on the sign of the invariant $(\kappa/hH - 1/4H^2)$. For $0 < h < 4\kappa H$ it is positive, and the solutions of (3.12) oscillate in the vertical. These solutions are called internal modes. For h outside this range the solution varies exponentially with z and is called a trapped or evanescent mode.

At $z = 0$ we require the vertical velocity to vanish. From (3.5) it follows that

$$\frac{d\hat{Z}}{dz} - \left(\frac{\kappa}{H} - \frac{1}{2H} \right) \hat{Z} = 0 \quad \text{at} \quad z = 0. \tag{3.13}$$

Two different upper boundary conditions are applied. For trapped modes, we require the energy density to decay with height. For internal modes, we impose a lid condition.

External modes

We first study the vertically trapped solutions. Defining $n^2 = |\kappa/hH - 1/4H^2|$, the solution is

$$\hat{Z} = Ae^{nz} + Be^{-nz}$$

and, in order for the energy density to remain bounded, we must have $A = 0$. The lower boundary condition (3.13) is then satisfied only if $h = \gamma H$. For typical parameter values, $H \approx 8\,\text{km}$, so that the equivalent depth is about 11 km. Z, and therefore u, v and $p/\bar{\rho}$ all grow exponentially with z but the energy density decays as $\exp[-(1 - 2\kappa)z/H]$. From (3.5) it follows that the vertical velocity vanishes identically for this solution. It is the same solution that we found in §3.1 by assuming $w \equiv 0$ ab initio. We call solutions with this vertical structure external modes.

Internal modes

Let us now assume $0 < h < 4\kappa H$ so that the invariant of (3.12) is positive. Again defining $n^2 = |\kappa/hH - 1/4H^2|$, the solution is

$$\hat{Z} = Ae^{inz} + Be^{-inz}.$$

The energy density is now independent of z so we require another upper boundary condition. We impose a lid condition $w = 0$ at some particular height z_T. This requires $\sin nz_T = 0$, so that $nz_T = \ell\pi$, which gives a discrete spectrum of solutions

$$h_\ell = \frac{4\kappa H}{1 + (2\ell\pi H/z_T)^2} \quad \text{for} \quad \ell = 1, 2, 3, \ldots$$

In fact, since the height z_T is arbitrary, we have a continuous spectrum of equivalent depths covering the range $0 < h < 4\kappa H$. These internal modes have sinusoidal variation of $\hat{Z}$ in the vertical. Therefore, u, v and $p/\bar{\rho}$ all grow exponentially with z as $\exp[z/2H]$ and the energy density is constant with height.

Appendix 4

Richardson's forecast factory: the $64 000 question

Reprinted from *The Meteorological Magazine*, **122**, 69–70.
(*March 1993*)

Lewis Fry Richardson served as a driver for the Friends Ambulance Unit in the Champagne district of France from September 1916 until the Unit was dissolved in January 1919 following the cessation of hostilities. For much of this time he worked near the front line, and during the Battle of Champagne in April 1917 he came under heavy shelling (Ashford, 1985). It is a source of wonder that in such appallingly inhuman conditions he had the buoyancy of spirit to carry out one of the most remarkable and prodigious calculational feats ever accomplished. During the intervals between transporting wounded soldiers back from the front he worked out by manual computation the changes in the pressure and wind at two points, starting from an analysis of the condition of the atmosphere at 0700 UTC on 20 May 1910. Richardson described his method of solving the equations of atmospheric motion and his sample forecast in what has become the most famous book in meteorology, his *Weather Prediction by Numerical Process* (Richardson, 1922). The unrealistic values which he obtained are a result of inadequacies and imbalances in the initial data, and do not reflect any flaw in his method, which is essentially the way numerical forecasts are produced today.

How long did it take Richardson to make his forecast? And how many people would be required to put the method to practical use? The answers to these two questions are contained in §11/2 of his book, but are expressed in a manner which has led to some confusion. On p. 219 under the heading 'The Speed and Organization of Computing' Richardson wrote

> It took me the best part of six weeks to draw up the computing forms and to work out the new distribution in two vertical columns for the first time. My office was a heap of hay in a cold rest billet. With practice the work of an average computer might go perhaps ten times faster. If the time-step were 3 hours, then 32 individuals could just compute two points so as to keep pace with the weather.

Could Richardson really have completed his task in six weeks? Given that 32 computers working at ten times his speed would require 3 hours for the job, he himself must have taken some 960 hours – that is 40 days or 'the best part of six weeks' working flat-out at 24 hours a day! At a civilized 40-hour week the forecast would have extended over six

259

months. It is more likely that Richardson spent perhaps ten hours per week at his chore and that it occupied him for about two years, the greater part of his stay in France.

Now to the question of the resources required to realize Richardson's dream of practical forecasting. Quoting again from p. 219 of the book:

> If the co-ordinate chequer were 200 km square in plan, there would be 3200 columns on the complete map of the globe. In the tropics the weather is often foreknown, so that we may say 2000 active columns. So that $32 \times 2000 = 64\,000$ computers would be needed to race the weather for the whole globe. That is a staggering figure.

It is indeed staggering, when we recall that these 'computers' were living, feeling beings, not senseless silicon chips. Richardson proposed taking 128 chequers or grid-boxes around each parallel and 100 between the poles. This gives a grid cell which is roughly a square of side 200 km at 50° North and South. He outlined a scheme for reducing the number of chequers towards the poles but made no allowance for that in the above reckoning. His claim that 3200 columns or chequers would cover the globe has been questioned by Sydney Chapman in his Introduction to the Dover Edition of *Weather Prediction by Numerical Process*:

> As to Richardson's estimates of the time and cost of full application of his methods, he made an uncharacteristic error in giving 3200 as the number of squares ... to cover the globe. His number is only a quarter of the true value, so that his required staff and his cost estimate must be quadrupled.

So, Chapman's estimate of the staff required is $4 \times 64\,000 = 256\,000$. However, this is not entirely correct. The envisaged computational grid would indeed have required $128 \times 100 = 12\,800$ chequers for global coverage – four times the value stated by Richardson. But Richardson considered the grid-boxes in pairs, one for mass and one for momentum, and it was such a pair for which he made his sample forecast and upon which he based his estimates. Thus, 6400 pairs of chequers would cover the globe and, with 32 people working on each pair, a total horde of 204 800 would be involved in a bid to race the weather for the whole globe. That is a stupendous figure!

So where did Richardson come by the figure of 3200 chequers to cover the globe? The error is inescapable but is not, I believe, a numerical slip. Richardson intimated that the weather in the tropics was sufficiently steady for variations to be neglected. But in such a case the global forecasting problem falls neatly into two parts and it is natural to consider each hemisphere separately. The Northern hemisphere can be covered by 3200 *pairs* of columns. Assuming with Richardson that the values at 1200 pairs may be prescribed and assigning 32 individuals to each of the remaining pairs, one finds that $32 \times 2000 = 64\,000$ souls are needed to race the weather *for the extra-tropical Northern hemisphere.*

If this is what Richardson intended, his 'uncharacteristic error' was not an arithmetical howler but a lapse of expositional precision. For his staggering figure of 64 000 is clearly stated to refer to *the whole globe.* Later in the paragraph he speaks of a forecast-factory for the whole globe (in fact, the word 'globe' occurs five times on p. 219). In his wonderful fantasy of a theatre full of computers, the tropics 'in the upper circle' are treated on an equal footing with the temperate and frigid zones. Given that Richardson's assumption of constancy of tropical weather was over-optimistic, a full complement of 32 computers for each pair of columns in his *forecast-factory for the whole globe* would have provided work for 204 800 people.

Even this vast multitude could compute the weather only as fast as it was evolving. To obtain useful and timely predictions, the calculations would need to go several times faster than the atmosphere. Allowing for a speed-up factor of five, the establishment of a 'practical' forecast-factory would have reduced the ranks of the unemployed by over a million.

References

Abbe, C. (1901). The physical basis of long-range weather forecasts. *Mon. Weather Rev.*, **29**, 551–61.

Antoniou, A. (1993). *Digital Filters: Analysis, Design and Applications*, 2nd edn., McGraw-Hill.

Arakawa, A. and Lamb, V. R. (1977). Computational design of the basic dynamical processes of the UCLA general circulation model. In *General Circulation Models of the Atmopshere*, ed. J. Chang. Academic Press, pp. 173–265.

Árnason, G. (1952). A baroclinic model of the atmosphere applicable to the problem of numerical forecasting in three dimensions. *Tellus*, **4**, 356–73.

Árnason, G. (1958). A convergent method for solving the balance equation. *J. Meteor.*, **15**, 220–5.

Ashford, O. M. (1985). *Prophet—or Professor? The Life and Work of Lewis Fry Richardson.* Adam Hilger.

Ashford, O. M. (2001). Lost and found. Letter in *Weather*, **56**, 114.

Asselin, R. (1972). Frequency filter for time integrations. *Mon. Weather Rev.*, **100**, 487–90.

Bachelor, G. F. (1950). The application of the similarity theory of turbulence to atmospheric diffusion. *Q. J. Roy. Meteor. Soc.*, **76**, 133–46.

Baer, F. (1977). Adjustment of initial conditions required to suppress gravity oscillations in nonlinear flows. *Beit. Phys. Atmos.*, **50**, 350–66.

Baer, F. and Tribbia, J. (1977). On complete filtering of gravity modes through non-linear initialization. *Mon. Weather Rev.*, **105**, 1536–9.

Bengtsson, L., Hodges, K. I. and Froude, L. S. R. (2005). Global observations and forecast skill. *Tellus*, **A-57**, 515–27.

Benjamin, S. G., Dévényi, D., Weygandt, S. S. *et al.* (2004). An hourly assimilation-forecast cycle: the RUC. *Mon. Weather Rev.*, **132**, 495–518.

Benwell, G. R. R., Gadd, A. J., Keers, J. F. *et al.* (1971). The Bushby–Timpson 10-level model on a fine mesh. Met Office Sci. Paper No. 32 (Met.O.836), HMSO.

Bergthorsson, P. and Döös, B. (1955). Numerical weather map analysis. *Tellus*, **7**, 329–40.

Birchfield, G. E. (1956). Summary of work being done in the field of numerical weather prediction. *CBI #36*, Box 1, Folder 1. Available from Charles Babbage Institute, University of Minnesota.

Bjerknes, V. (1904). Das Problem der Wettervorhersage, betrachtet vom Standpunkte der Mechanik und der Physik. *Meteor. Zeit.*, **21**, 1–7. Translation by Y. Mintz: The problem of weather forecasting as a problem in mechanics and physics. Los Angeles, 1954. Reprinted in Shapiro and Grønås (1999), pp. 1–4.

Bjerknes, V. (1910). Synoptical representation of atmospheric motions. *Q. J. Roy. Meteor. Soc.*, **36**, 267–86.

Bjerknes, V. (1914a). Meteorology as an exact science. Translation of *Die Meteorologie als exacte Wissenschaft*. Antrittsvorlesung gehalten am 8. Jan. 1913 in der Aula der Universität Leipzig. *Mon. Weather Rev.*, **42**, 11–14.

Bjerknes, V. (1914b). *Veröffentlichungen des Geophysikalischen Instituts der Universität Leipzig*. Herausgegeben von dessen Direktor V. Bjerknes. Erste Serie: *Synoptische Darstellungen atmosphärischer Zustände*. Jahrgang 1910, Heft 3. Zustand der Atmosphäre über Europa am 18., 19. und 20. Mai 1910. Leipzig, 1914.

Bjerknes, V. (1921). On the dynamics of the circular vortex. *Geofysisk. Publ.*, **2**, No. 4.

Bjerknes, V. and Sandström, J.W. (1910). *Dynamic Meteorology and Hydrography. Part I: Statics*. Carnegie Institution, Publication 88 (Part I).

Bjerknes, V., Hesselberg, Th. and Devik, O. (1911). *Dynamic Meteorology and Hydrography. Part II: Kinematics*. Carnegie Institution, Publication 88 (Part II).

Bolin, B. (1956). An improved barotropic model and some aspects of using the balance equation for three-dimensional flow. *Tellus*, **8**, 61–75.

Brathseth, A. (1986). Statistical interpolation by means of successive corrections. *Tellus*, **38A**, 439–47.

Brillouin, M. (1932). Marées dynamiques. Les latitudes critiques. *Compt. Rend. Acad. Sci.*, Paris, **194**, 801–4.

Brunt, D. (1954). Obituary of Richardson. *Q. J. Roy. Meteor. Soc.*, **80**, 127–8.

Bubnová, R., Hello, G., Bénard, P. *et al.* (1995). Integration of the fully elastic equations cast in the hydrostatic pressure terrain-following coordinate in the framework of the ARPEGE/Aladin NWP system. *Mon. Weather Rev.*, **123**, 515–35.

Buizza, R., Miller, M. J. and Palmer, T. N. (1999). Stochastic simulation of model uncertainties in the ECMWF ensemble prediction system. *Q. J. Roy. Meteor. Soc.*, **125**, 2887–908.

Bushby, F. H. (1986). A history of numerical weather prediction. In *Short- and Medium-Range Weather Prediction*, Collection of papers presented at the WMO/IUGG NWP Symposium, Tokyo, 4–8 August, 1986. Special issue of *J. Met. Soc. Japan*, 1–10.

Bushby, F. H. and Timpson, M. S. (1967). A 10-level atmospheric model and frontal rain. *Q. J. Roy. Meteor. Soc.*, **93**, 1–17.

Bushby, F. H. and Whitelam, C. J. (1961). A three-parameter model of the atmosphere suitable for numerical investigation. *Q. J. Roy. Meteor. Soc.*, **87**, 380–92.

Cartwright, D. E. (1999). *Tides. A Scientific History*. Cambridge University Press.

Chapman, S. (1965). Introduction to Dover edition of *Weather Prediction by Numerical Process* (See Richardson, 1922).

Chapman, S. and Lindzen, R. S. (1970). *Atmospheric Tides, Thermal and Gravitational*. Reidel Publ. Co.

Charney, J. G. (1947). The dynamics of long waves in a baroclinic westerly current. *J. Meteor.*, **4**, 135–62.

Charney, J. G. (1948). On the scale of atmospheric motions. *Geofysisk. Publ.*, **17**, No. 2.

Charney, J. G. (1949). On a physical basis for numerical prediction of large-scale motions in the atmosphere. *J. Meteor.*, **6**, 371–85.

Charney, J. G. (1950). Progress in dynamic meteorology. *Bull. Amer. Met. Soc.*, **31**, 231–6.

Charney, J. G. (1951). Dynamic forecasting by numerical process. In *Compendium of Meteorology*. American Meteorological Society, pp. 470–82.

Charney, J. G. (1955). The use of the primitive equations of motion in numerical prediction. *Tellus*, **7**, 22–6

Charney, J. G. (1962). Integration of the primitive and balance equations. Proc. Intern. Symp. Numerical Weather Prediction. Japan. Meteorology Agency, Tokyo, pp. 131–6.

Charney, J. G. and Eliassen, A. (1949). A numerical method for predicting the perturbations of the middle latitude westerlies. *Tellus*, **1**, 38–54.

Charney, J. G., Fjørtoft, R. and von Neumann, J. (1950). Numerical integration of the barotropic vorticity equation. *Tellus*, **2**, 237–54.

Charney, J. G. and Phillips, N. A. (1953). Numerical integration of the quasi-geostrophic equations for barotropic and simple baroclinic flows. *J. Meteor.*, **10**, 71–99.

Courant, R., Friedrichs, K. O. and Lewy, H. (1928). Über die partiellien Differenzengleichungen der mathematischen Physik. *Math. Annalen*, **100**, 32–74.

Courtier, P., Thépaut, J. N. and Hollingsworth, A. (1994). A strategy for operational implementation of 4D-Var, using an incremental approach. *Q. J. Roy. Meteor. Soc.*, **120**, 1367–87.

Cressman, G. P. (1958). Barotropic divergence and very long atmospheric waves. *Mon. Weather Rev.*, **86**, 293–7.

Cressman, G. P. (1959). An operational objective analysis system. *Mon. Weather Rev.*, **87**, 367–74.

Cressman, G. P. (1996). The origin and rise of numerical weather prediction. In Fleming (1996) pp. 21–39.

Cullen, M. J. P. (1993). The unified forecast/climate model. *Meteorol. Mag.*, **122**, 81–94.

Daley, R. (1991). *Atmospheric Data Analysis*. Cambridge University Press.

Davies, H. C. (1976). A lateral boundary formulation for multi-level prediction models. *Q. J. Roy. Meteor. Soc.*, **102**, 405–18.

Dickinson, R. and Williamson, D. (1972). Free oscillations of a discrete stratified fluid with application to numerical weather prediction. *J. Atmos. Sci.*, **29**, 623–40.

Dikii, L. A. (1965). The terrestrial atmosphere as an oscillating system. *Izvestiya, Atmos. and Oceanic Phys.*, **1**, 469–89.

Dines, W. H. (1909). Contributions to the investigation of the upper air. Report on apparatus and methods used at Pyrton Hill. Met. Office, MO 202, pp. 15–38. (Reprinted in Dines, 1931).

Dines, W. H. (1912). The free atmosphere in the region of the British Isles. Third Report. The calibration of the balloon meteorograph and the reading of the traces. Met. Office Geophys. Memoirs, Vol. 1, No. 6, 145–152 (M.O. No. 210f).

Dines, W. H. (1919). The characteristics of the free atmosphere. Met. Office, Geophys. Memoirs, No 13, pp. 47–76. (Reprinted in Dines, 1931).

Dines, W. H. (1931). *Collected Scientific Papers*. Royal Meteor. Soc.

Doms, G. and Schättler, U. (1998). The nonhydrostatic limited-area model LM (Lokal-Modell) of DWD, Part I: Scientific documentation. German Weather Service, Research Department.

Drazin, P. G. (1993). See LFR I.

Durran, D. R. (1999). *Numerical Methods for Wave Equations in Geophysical Fluid Dynamics*. Springer.

DWD (1956). Symposium über Numerische Wettervorhersage in Frankfurt am Main. *Berichte des Deutschen Wetterdienstes*, Nr. 38, Band 5.

Eady, E. T. (1949). Long waves and cyclone waves. *Tellus*, **1**, 33–52.

Eady, E. T. (1952). Note on weather computing and the so-called $2\frac{1}{2}$-dimensional model. *Tellus*, **4**, 157–67.

Eliassen, A. (1949). The quasi-static equations of motion with pressure as independent variable. *Geofysisk. Publ.*, **17**, No. 3.

Eliassen, A. (1952). Simplified models of the atmosphere, designed for the purpose of numerical weather prediction. *Tellus*, **4**, 145–56.

Eliassen, A. (1984). Geostrophy. Symons memorial lecture, May, 1983. *Q. J. Roy. Meteor. Soc.*, **110**, 1–12.

Eliassen, A. (1999). Vilhelm Bjerknes' early studies of atmospheric motions and their connection with the cyclone model of the Bergen School. In Shapiro and Grønås (1999), pp. 5–13.

Exner, F. M. (1908). Über eine erste Annäherung zur Vorausberechnung synoptischer Wetterkarten. *Meteor. Zeit.*, **25**, 57–67. Translated from German as: *A first approach towards calculating synoptic forecast charts*, with a biographical note on Exner by L. Shields and an introduction by P. Lynch. Historical Note No. 1, Met Éireann, 1995.

Exner, F. M. (1917). *Dynamische Meteorologie*. B. D. Teubner.

Exner, F. M. (1923). Review of "Weather Prediction by Numerical Process". *Meteor. Zeit.*, **40**, 189–90.

Fairgrieve, M. A. (1913). On the relation between the velocity of the gradient wind and that of the observed wind. Met. Office Geophys. Memoirs, Vol. 1, No. 9, 187–208 (M.O. No. 210i).

Ferrel, W. (1859). The motions of fluids and solids relative to the Earth's surface. *Math. Monthly*, **1**, 140–7, 210–16, 300–7, 366–72, 397–406.

Fjørtoft, R. (1952). On a numerical method of integrating the barotropic vorticity equation. *Tellus*, **4**, 179–94.

Flattery, T. W. (1967). *Hough Functions*. Tech. Rep. No. 21 to the National Science Foundation, Dept. of Geophysical Sciences, University of Chigago.

Fleming, J. R. ed. (1996). *Historical Essays on Meteorology, 1919–1995*. American Meteorological Society.

Fortak, H. (2001). Felix Maria Exner und die österreichische Schule der Meteorologie. In Hammerl *et al.*, 2001, pp. 354–386.

Fox, L. (1993). Numerical Analysis. In Richardson's *Collected Papers* (see LFR I), pp. 36–44.

Friedman, R. M. (1989). *Appropriating the Weather: Vilhelm Bjerknes and the Construction of a Modern Meteorology*. Cornell Univ. Press.

Gandin, L. (1963). *Objective Analysis of Meteorological Fields* (Leningrad: Gidromet). English translation (Jerusalem: Israel Program for Scientific Translation) 1965.

Geisler, J. E. and Dickinson, R. E. (1976). The five-day wave on a sphere with realistic zonal winds. *J. Atmos. Sci.*, **33**, 632–41.

Gilchrist, B. and Cressman, G. (1954). An experiment in objective analysis. *Tellus*, **6**, 309–18.

Gleick, J. (1992). *Genius: Richard Feynman and Modern Physics*. Little, Brown & Co.

Gold, E. (1920). Aids to forecasting; types of pressure distribution, with notes and tables for the fourteen years 1905–1918. *Met. Office Geophys. Mem.*, no. 16.

Gold, E. (1954). Lewis Fry Richardson. Obituary notices of Fellows of the Royal Society, London, **9**, 217–35.

Goldstine, H. H. (1972). *The Computer from Pascal to von Neumann*. Reprinted with new preface, Princeton University Press, 1993.

Haltiner, G. J. and Williams, R. T. (1980). *Numerical Prediction and Dynamic Meteorology*. John Wiley and Sons.

Hammerl, C., Lenhardt, W., Steinacker, R, *et al.* (2001). *Die Zentralanstalt für Meteorologie und Geodynamik 1851–2001. 150 Jahre Meteorologie und Geophysik in Österreich*. Leykam Buchverlags GmbH.

Hamming, R. W. (1989). *Digital Filters*. Prentice-Hall International.

Haurwitz, B. (1940). The motion of atmospheric disturbances on the spherical earth. *J. Marine Res.*, **3**, 254–67.

Haurwitz, B. (1941). *Dynamic Meteorology.* McGraw-Hill Book Company.

Heims, S. J. (1980). *John Von Neumann and Norbert Wiener: From Mathematics to the Technologies of Life and Death.* MIT Press.

Hergesell, H. (1905). Begrüssungsworte. Fourth conference of the International Commission for Scientific Aeronautics, St Petersburg, 1904. H. Hergesell, ed., 28–35.

Hergesell, H. (1913). *Veröffentlichungen der Internationalen Kommission für wissenschaftliche Luftschiffahrt.* Jahrgang 1910, Heft 5: Herausgegeben von Prof. Dr. H. Hergesell. *Beobachtungen mit bemannten, unbemannten Ballons und Drachen sowie auf Berg- und Wolkenstationen vom 18.–20. Mai 1910.* Straßburg, DuMont Schauberg.

Herrlin, O. (1956). Numerical Forecasting at the Swedish Military Meteorological Office in 1954–1956. In DWD, 1956, pp. 53–55.

Hess, G. D. (1995). An introduction to Lewis Fry Richardson and his mathematical theory of war and peace. *Conflict Management and Peace Science*, **14**, 77–113.

Hinds, M. K. (1981). Computer story. *Meteorol. Mag.*, **110**, 69–81.

Hinkelmann, K. (1951). Der Mechanismus des meteorologischen Lärmes. *Tellus*, **3**, 285–96. Translation: The mechanism of meteorological noise. NCAR/TN-203+STR, National Center for Atmospheric Research, 1983.

Hinkelmann, K. (1959). Ein numerisches Experiment mit den primitiven Gleichungen. In *The Atmosphere and the Sea in Motion.* Rossby Memorial Volume, Rockerfeller Institute Press, 486–500.

Hinkelmann, K. (1969). Primitive Equations. In *Lectures in numerical short-range weather prediction*. Regional Training Seminar, Moscow. WMO No. 297, pp. 306–75.

Hoinka, K. P. (1997). The tropopause: discovery, definition and demarcation. *Meteor. Zeit.*, N.F., **6**, 281–303.

Holl, P. (1970). Die Vollständigkeit des Orthogonalsystems der Hough-Funktionen. *Nachr. Akad. Wissens. Göttingen*, Math-Phys. Kl. No.7, 159–68.

Hollingsworth, A. (1994). Validation and diagnosis of atmospheric models. *Dynam. Atmos. Sci.*, **20**, 227–46.

Holton, J. R. (1975). *The Dynamic Meteorology of the Stratosphere and Mesosphere.* Met. Monogr., Vol. 15, No. 37. American Meteorological Society.

Holton, J. R. (2004). *An Introduction to Dynamic Meteorology.* 4th edn. Elsevier Academic Press.

Hough, S. S. (1898). On the application of harmonic analysis to the dynamical theory of the tides: Part II: on the general integration of Laplace's dynamical equations. *Phil. Trans. Roy. Soc., London*, A, **191**, 139–85.

Huang, X.-Y. and Lynch, P. (1993). Diabatic digital filtering initialization: application to the HIRLAM model. *Mon. Weather Rev.*, **121**, 589–603.

Hunt, J. C. R. (1993). A general introduction to the life and work of L. F. Richardson. In *Collected Papers of Lewis Fry Richardson*, pp. 1–27.

Hunt, J. C. R. (1998). Lewis Fry Richardson and his contributions to mathematics, meteorology, and models of conflict. *Ann. Rev. Fluid Mech.*, **30**, xiii–xxxvi.

Janssen, P. (2004). *The Interaction of Ocean Waves and Wind.* Cambridge University Press.

Jung, T., Klinker, E. and Uppala, S. (2003). Reanalysis and reforecast of three major European storms of the 20th century using the ECMWF forecasting system. ERA-40 Project Report Series No. 10 (ECMWF).

Kållberg, P., Berrisford, P., Hoskins, B. *et al.* (2005). ERA-40 Atlas. ERA-40 Project Report Series No. 19 (ECMWF).

Kalnay, E. (2003). *Atmospheric Modeling, Data Assimilation and Predictability.* Cambridge University Press.

Kasahara, A. (1976). Normal modes of ultralong waves in the atmosphere. *Mon. Weather Rev.*, **104**, 669–90.

Kasahara, A. (2000). On the origin of cumulus parameterization for numerical prediction models. In *General Circulation Model Development*, ed. D. A. Randall, Academic Press, pp. 199–224.

Kasahara, A. and Qian, J.-H. (2000). Normal modes of a global nonhydrostatic atmospheric model. *Mon. Weather Rev.*, **128**, 3357–75.

Kasahara, A. and Washington, W. M. (1967). NCAR global general circulation model of the atmosphere. *Mon. Weather Rev.*, **95**, 389–402.

Khrgian, A. Kh. (1959). *Meteorology. A Historical Survey.* Vol. I. Translated from the Russian by Ron Hardin. Israel Programme for Scientific Translations (IPST Cat. No. 5565), 1970.

Kolmogorov, A. N. (1941). The local structure of turbulence in incompressible viscous fluid for very large Reynolds numbers. *Dokl. Akad. Nauk. SSSR*, **30**, 301–5. English translation in "Turbulence and stochastic processes: Kolmogorov's ideas 50 years on", *Proc. R. Soc. London*, A 434 (1991), 9–17.

Körner, T. W. (1996). *The Pleasures of Counting.* Cambridge University Press.

Kuo, H.-L. (1949). Dynamic instability of two-dimensional nondivergent flow in a barotropic atmosphere. *J. Meteor.*, **6**, 105–22.

Kurihara, Y. (1965). Numerical integration of the primitive equations on a spherical grid. *Mon. Weather Rev.*, **93**, 399–415.

Kutzbach, G. (1979). *The Thermal Theory of Cyclones. A History of Meteorological Thought in the Nineteenth Century.* Historical Monograph Series, American Meteorological Society.

Lamb, H. (1932). *Hydrodynamics*, 6th edn., Cambridge University Press.

Lancaster-Browne, P. (1985). *Halley and His Comet.* Blandford Press.

Leith, C. E. (1980). Nonlinear normal mode initialization and quasi-geostrophic theory. *J. Atmos. Sci.*, **37**, 958–68.

Lewis, J. M. (1998). Clarifying the dynamics of the general circulation: Phillips's 1956 experiment. *Bull. Amer. Met. Soc.*, **79**, 39–60.

LFR I (1993). *Collected Papers of Lewis Fry Richardson,* Vol. I. *Meteorology and Numerical Analysis.* General editor P. G. Drazin, Cambridge University Press.

LFR II (1993). *Collected Papers of Lewis Fry Richardson,* Vol. II. *Quantitative Psychology and Studies of Conflict.* General editor I. Sutherland, Cambridge University Press.

Lighthill, J. (1978). Mathematics and the physical environment. In *Newer Uses of Mathematics*, ed. J. Lighthill. Penguin Books, pp. 13–107.

Lin, C. A., Laprise, R. and Richie, H. ed. (1997). *Numerical Methods in Atmospheric and Oceanic Modelling. The André J. Robert Memorial Volume.* Canadian Meteorological and Oceanographic Society.

Lindzen, R. S. (1968). Rossby waves with negative equivalent depth — comments on a note by G. A. Corby. *Q. J. Roy. Meteor. Soc.*, **94**, 402.

Lindzen, R. S., Lorenz, E. N. and Platzman, G. W. eds. (1990). *The Atmosphere—A Challenge. The Science of Jule Gregory Charney,* American Meteorological Society.

Loehrer, S. M. and Johnson, R. H. (1995). Surface presure and precipitation life cycle characteristics of PRE-STORM mesoscale convective systems. *J. Atmos. Sci.*, **123**, 600–21.

Longuet-Higgins, M. S. (1968). The eigenfunctions of Laplace's tidal equations over a sphere. *Phil. Trans. Roy. Soc., London*, **A262**, 511–607.

Lorenc, A. (1986). Analysis methods for numerical weather prediction. *Q. J. Roy. Meteor. Soc.*, **112**, 1177–94.

Lorenz, E. N. (1963). Deterministic nonperiodic flow. *J. Atmos. Sci.*, **20**, 130–41.

Lorenz, E. N. (1967). *The Nature and Theory of the General Circulation of the Atmosphere.* World Meteor. Org., WMO No. 218 TP.

Lorenz, E. N. (1980). Attractor sets and quasi-geostrophic equilibrium. *J. Atmos. Sci.*, **37**, 1685–99.

Lorenz, E. N. (1993). *The Essence of Chaos.* University of Washington Press.

Lynch, P. (1979). Baroclinic instability of ultra-long waves modelled by the planetary geostrophic equations. *Geophys. Astrophs. Fluid Dynam.*, **13**, 107–24.

Lynch, P. (1989). Partitioning the wind in a limited domain. *Mon. Weather Rev.*, **117**, 1492–500.

Lynch, P. (1992). Richardson's barotropic forecast: a reappraisal. *Bull. Amer. Met. Soc.*, **73**, 35–47.

Lynch, P. (1993). Richardson's forecast factory: the $ 64 000 question. *Met. Mag.*, **122**, 69–70.

Lynch, P. (1997). The Dolph–Chebyshev window: a simple optimal filter. *Mon. Weather Rev.*, **125**, 655–60.

Lynch, P. (1999). Richardson's marvellous forecast. In *The Life Cycles of Extratropical Cyclones*, M. A. Shapiro and S. Grønås, eds., American Meteorological Society, pp. 61–73.

Lynch, P. (2001). Max Margules and his tendency equation. Historical Note No. 5, Met Éireann.

Lynch, P. (2002). The swinging spring: a simple model for atmospheric balance. In Norbury and Roulstone (2002b), pp. 64–108.

Lynch, P. (2003a). Introduction to initialization. In Swinbank *et al.* (2003), pp. 97–111.

Lynch, P. (2003b). Digital filter initialization. In Swinbank *et al.* (2003), pp. 113–26.

Lynch, P. (2003c). Margules' tendency equation and Richardson's forecast. *Weather*, **58**, 186–93.

Lynch, P. and Huang, X.-Y. (1992). Initialization of the HIRLAM model using a digital filter. *Mon. Weather Rev.*, **120**, 1019–34.

Lynch, P., Giard, D. and Ivanovici, V. (1997). Improving the efficiency of a digital filtering scheme. *Mon. Weather Rev.*, **125**, 1976–82.

McAdie, A. (1923). Book Review. *Geograph. Rev.*, **13**, 324–5.

McDonald, A. (1986). A semi-Lagrangian and semi-implicit two time-level integration scheme. *Mon. Weather Rev.*, **114**, 824–30.

McDonald, A. (1997). Lateral boundary conditions for operational regional forecast models; a review. Hirlam Tech. Rep. No. 32, Met Éireann.

McDonald, A. and Bates, J. R. (1989). Semi-Lagrangian integration of a grid-point shallow water model on the sphere. *Mon. Weather Rev.*, **117**, 130–7.

McIntyre, M. E. (2003). Balanced flow. In *Encyclopedia of Atmospheric Sciences*, eds. J. R. Holton, J. Pyle and J. A. Curry. 5 vols, Acad. Press.

Machenhauer, B. (1977). On the dynamics of gravity oscillations in a shallow water model, with applications to normal mode initialization. *Beit. Phys. Atmos.*, **50**, 253–71.

Macrae, N. (1999). *John von Neumann*, American Mathematical Society. Originally published by Pantheon Books, 1992.

Madden, R. A. (1979). Observations of large-scale traveling Rossby waves. *Rev. Geophys. Space Phys.*, **17**, 1935–49.

Madden, R. A. and Julian, P. A. (1972). Further evidence of global-scale 5-day pressure waves. *J. Atmos. Sci.*, **29**, 1464–9.

Majewski, D., Liermann, D., Prohl, P. *et al.* (2002). The operational global icosahedral-hexagonal gridpoint model GME: description and high-resolution tests. *Mon. Weather Rev.*, **130**, 319–38.

Mandelbrot, B. B. (1967). How long is the coast of Britain? Statistical self-similarity and fractional dimension. *Science*, **156**, 636–8.

Mandelbrot, B. B. (1982). *The Fractal Geometry of Nature.* W. H. Freeman & Co.

Marchuk, G. I. (1982). *Methods of Numerical Mathematics*, 2nd edn. Springer-Verlag.

Margules, M. (1893). Luftbewegungen in einer rotierenden Sphäroidschale. *Sitzungsberichte der Kaiserliche Akad. Wiss. Wien*, IIA, **102**, 11–56.

Margules, M. (1904). Über die Beziehung zwischen Barometerschwankungen und Kontinuitätsgleichung. *Ludwig Boltzmann Festschrift.* J A Barth. [*On the relationship between barometric variations and the continuity equation.* Translation in Lynch, 2001.]

MC-1911: Meteorological Committee (1911). Sixth Annual Report of the Meteorological Committee to the Lords Commissioners of His Majesty's Treasury, for the year ended 31st March 1911. HMSO.

Mesinger, F. (2000). Limited area modelling: beginnings, state of the art, outlook. In *50th Anniversary of Numerical Weather Prediction. Commemorative Symposium*, ed. A. Spekat Deutsche Meteorologische Gesellschaft, pp. 91–118.

Mesinger, F. and Arakawa, A. (1976). *Numerical Methods used in Atmospheric Models.* Vol 1. WMO/ICSU Joint Organizing Committee, GARP Publication Series No. 17.

Mischaikow, K. and Mrozek, M. (1995). Chaos in the Lorenz equations: a computer-assisted proof. *Bull. Amer. Math. Soc.*, **32**, 66–72.

Miyakoda, K. and Moyer, R. (1968). A method of initialization for dynamical weather forecasting. *Tellus*, **20**, 115–28.

Morse, P. M. and Feshbach, H. (1953). *Methods of Theoretical Physics, Part I.* McGraw-Hill.

NCUACS (1993). Catalogue of the papers and correspondence of Lewis Fry Richardson deposited in Cambridge University Library. Compiled by T. E. Powell and P. Harper, National Cataloguing Unit for the Archives of Contemporary Science, University of Bath, 1993.

Nebeker, F. (1995). *Calculating the Weather: Meteorology in the 20th Century.* Academic Press.

Nitta, T. and Hovermale, J. (1969). A technique of objective analysis and initialization for the primitive forecast equations. *Mon. Weather Rev.*, **97**, 652–8.

Norbury, J. and Roulstone, I. (2002a). *Large-Scale Atmosphere-Ocean Dynamics.* Vol. 1: *Analytical Methods and Numerical Models.* Cambridge University Press.

Norbury, J. and Roulstone, I. (2002b). *Large-Scale Atmosphere-Ocean Dynamics.* Vol. 2: *Geometric Methods and Models.* Cambridge University Press.

Ohring, G. (1964). A most surprising discovery. *Bull. Amer. Met. Soc.*, **45**, 12–14.

Oliger, J. and Sundström, A. (1978). Theoretical and practical aspects of some initial boundary value problems in fluid dynamics. *SIAM J. Appl. Math.*, **35**, 419–46.

Oppenheim, A. V. and Schafer, R. W. (1989). *Discrete-Time Signal Processing.* Prentice-Hall International, Inc.

Palmer, T. N. (1993). Extended-range atmospheric prediction and the Lorenz model. *Bull. Amer. Met. Soc.*, **74**, 49–65.

Palmer, T. N., Shutts, G. J., Hagedorn, R. *et al.* (2005). Representing model uncertainty in weather and climate prediction. *Annual Review of Earth and Planetary Sciences, 33,* 163–93.

Panofsky, H. (1949). Objective weather-map analysis. *J. Appl. Meteor.*, **6**, 386–92.

Parrish, D. F. and Derber, J. D. (1992). The National Meteorological Center spectral statistical interpolation analysis system. *Mon. Weather Rev.*, **120**, 1747–63.

Persson, A. (2005a). Early operational numerical weather prediction outside the USA: an historical introduction. Part 1: Internationalism and engineering NWP in Sweden, 1952–69. *Meteor. Appl.*, **12**, 135–60.

Persson, A. (2005b). Early operational numerical weather prediction outside the USA: an historical introduction: Part II: Twenty countries around the world. *Meteor. Appl.*, **12**, 269–89.

Persson, A. (2005c). Early operational numerical weather prediction outside the USA: an historical introduction. Part III: Endurance and mathematics–British NWP, 1948–1965. *Meteor. Appl.*, **12**, 357–70.

Peterson, I. (1993). *Newton's Clock: Chaos in the Solar System.* W. H. Freeman & Co.

Petterssen, S. (2001). *Weathering the Storm. Sverre Petterssen, the D-Day Forecast, and the Rise of Modern Meteorology.* Fleming, J. R. ed. American Meteorological Society.

Phillips, N. A. (1951). A simple three dimensional model for the study of large-scale extra-tropical flow patterns. *J. Meteor.*, **8**, 381–94.

Phillips, N. A. (1953). A coordinate system having some special advantages for numerical forecasting. *J. Meteor.*, **14**, 184–5.

Phillips, N. A. (1956). The general circulation of the atmosphere: a numerical experiment. *Q. J. Roy. Meteor. Soc.*, **82**, 123–64.

Phillips, N. A. (1960). On the problem of initial data for the primitive equations. *Tellus*, **12**, 121–6

Phillips, N. A. (1963). Geostrophic motion. *Rev. Geophys.*, **1**, 123–76.

Phillips, N. A. (1966). The equations of motion for a shallow rotating atmosphere and the 'traditional approximation'. *J. Atmos. Sci.*, **23**, 626–8.

Phillips, N. A. (1973). Principles of large scale numerical weather prediction. In *Dynamic Meteorology,* ed. P. Morel. D. Reidel, pp. 1–96.

Phillips, N. A. (1990). The emergence of quasi-geostrophic theory. In Lindzen *et al.* (1990), pp. 177–206.

Phillips, N. A. (2000). The start of numerical weather prediction in the United States. In *50th Anniversary of Numerical Weather Prediction. Commemorative Symposium*, ed. A. Spekat. Deutsche Meteorologische Gesellschaft, pp. 13–28.

PIV (2005). *Plena Ilustrita Vortaro de Esperanto.* Sennacieca Asocio Tutmonda.

Platzman, G. W. (1967). A retrospective view of Richardson's book on weather prediction. *Bull. Amer. Met. Soc.*, **48**, 514–50.

Platzman, G. W. (1968). Richardson's weather prediction. *Bull. Amer. Met. Soc.*, **49**, 496–500.

Platzman, G. W. (1979). The ENIAC computations of 1950—gateway to numerical weather prediction. *Bull. Amer. Met. Soc.*, **60**, 302–12.

Platzman, G. W. (1990). Charney's Recollections. In Lindzen *et al.* (1990), pp. 11–82.

Poincaré, H. (1908). *Science and Method.* Reprinted by Dover Publications, 2003.

Press, W. H., Flannery, B. P., Teukolsky, S. A. *et al.* (1992). *Numerical Recipes in* FORTRAN, 2nd edition. Cambridge University Press.

Purdom, J. F. W. and Menzel, W. P. (1996). Evolution of satellite observations in the United States and their use in meteorology. In Fleming (1996), pp. 99–155.

Raymond, W. H. (1988). High-order low-pass implicit tangent filters for use in finite area calculations. *Mon. Weather Rev.*, **116**, 2132–41.

Recollections: see Platzman (1990).

Reid, C. (1976). *Courant in Göttingen and New York*. Springer-Verlag.

Reiser, H. (2000). The development of numerical weather prediction in the Deutscher Wetterdienst. In *50th Anniversary of Numerical Weather Prediction. Commemorative Symposium*. ed. A. Spekat. Deutsche Meteorologische Gesellschaft, pp. 51–78.

Richardson, L. F. (1910.) The approximate arithmetical solution by finite-differences of physical problems involving differential equations, with an application to the stresses in a masonry dam. *Phil. Trans. Roy. Soc.*, **A210**, 307–57 (see LFR I, pp.121–71).

Richardson, L. F. (1920). The supply of energy from and to atmospheric eddies. *Proc. Roy. Soc.*, **A97**, 354–73 (see LFR I, pp. 278–97).

Richardson, L. F. (1922). *Weather Prediction by Numerical Process*. Cambridge University Press. Reprinted by Dover Publications, 1965, with a new introduction by S. Chapman. Reprinted by Cambridge University Press, 2006, with a new introduction by Peter Lynch.

Richardson, L. F. (1925). How to solve differential equations approximately by arithmetic. *Math. Gaz.*, **12**, 415–21.

Richardson, L. F. (1926). Atmospheric diffusion shown on a distance-neighbour. *Proc. Roy. Soc.*, **A110**, 709–37.

Richardson, L. F. (1927). The deferred approach to the limit. Part I: Single lattice. *Phil. Trans. Roy. Soc., London*, **A226**, 299–349.

Richardson, L. F. (1951). Could an arms race end without fighting? *Nature*, **168**, 567.

Richardson, L. F. and Munday, R. E. (1926). The single-layer problem in the atmopshere and the height-integral of pressure. *Mem. Roy. Met. Soc.*, **1**, No. 2, 17–34.

Robert, A. (1979). The semi-implicit method. *Numerical Methods used in Atmospheric Models. Vol II*, Chapter 8. WMO/ICSU Joint Organizing Committee, GARP Publication Series No. 17, 417–37.

Rodgers, C. D. (1976). Evidence for the five-day wave in the upper stratosphere. *J. Atmos. Sci.*, **33**, 710–11.

Rossby, C.-G. (1945). On the propagation of frequencies and energy in certain types of oceanic and atmospheric waves. *J. Atmos. Sci.*, **2**, 187–204.

Rossby, C.-G. *et al.* (1939). Relations between variations in the intensity of the zonal circulation of the atmosphere and the displacements of the semipermanent centers of action. *J. Marine Res.*, **2**, 38–55.

Sadourny, R. (1975). The dynamics of finite difference models of the shallow water equations. *J. Atmos. Sci.*, **32**, 680–9.

Sanders, F. and Gyakum, J. R. (1980). Synoptic-dynamic climatology of the 'bomb'. *Mon. Weather Rev.*, **108**, 1589–606.

Santurette, P. and Georgiev, C. (2005). *Weather Analysis and Forecasting: Applying Satellite Water Vapor Imagery and Potential Vorticity Analysis*. Elsevier Academic Press.

Sasaki, Y. (1958). An objective analysis based on the variational method. *J. Met. Soc. Japan*, **36**, 77–88.

Sasaki, Y. (1970). Some basic formalisms in numerical variational analysis. *Mon. Weather Rev.*, **98**, 875–83.

Sawyer, J. S. and Bushby, F. H. (1953). A baroclinic model atmosphere suitable for numerical integration. *J. Meteor.*, **10**, 54–9.

Schmauss, A. (1911). *Deutsches Meteorologisches Jahrbuch für 1910*. Beobachtungen der Meteorologischen Stationen im Königreich Bayern im Jahre 1910.

Veröffentlichungen der Königlich Bayerischen Meteorologischen Centralstation. Herausgegeben durch deren Direktor August Schmauss, München, 1911.

Shapiro, M. A. and Grønås, S. eds. (1999). *The Life Cycles of Extratropical Cyclones*, American Meteorological Society.

Shaw, N. (1922). Review of *Weather Prediction by Numerical Process. Nature*, **110**, 762–5.

Shaw, N. (1932). *Manual of Meteorology*, Vol. 1. Cambridge University Press.

Sheppard, W. F. (1899). Central-difference formulae. *Proc. Lond. Math. Soc.*, Vol. XXXI, 449–88.

Shuman, F. G. (1957). Predictive consequences of certain physical inconsistencies in the geostrophic barotropic model. *Mon. Weather Rev.*, **85**, 229–34.

Shuman, F. G. (1989). History of numerical weather prediction at the National Meteorological Center. *Wea. & Forecast.*, **4**, 286–96.

Shuman, F. G. and Hovermale, J. (1968). An operational six-layer primitive equation model. *J. Appl. Meteor.*, **7**, 525–47.

Simmons, A. J. and Hollingsworth, A. (2002). Some aspects of the improvement in skill of numerical weather prediction. *Q. J. Roy. Meteor. Soc.*, **128**, 647–77.

Smagorinsky, J. (1958). On the numerical integration of the primitive equations of motion for baroclinic flow in a closed region. *Mon. Weather Rev.*, **86**, 457–66.

Smith, C. D. (1950). The destructive storm of November 25–27, 1950. *Mon. Weather Rev.*, **78**, 204–9.

Smith, W. L., Nagle, F. W., Hayden, C. M. *et al.* (1981). Vertical mass and moisture structure from TIROS-N. *Bull. Amer. Met. Soc.*, **62**, 388–93.

Spekat, A. ed. (2000). *50th Anniversary of Numerical Weather Prediction. Commemorative Symposium.* Deutsche Meteorologische Gesellschaft.

Stommel, H. M. (1985). Review of *Prophet—or Professor? The Life and Work of Lewis Fry Richardson* by O. M. Ashford. *Bull. Amer. Met. Soc.*, **66**, 1317.

Sutcliffe, R. C. (1947). A contribution to the problem of development. *Q. J. Roy. Meteor. Soc.*, **73**, 370–83.

Sutcliffe, R. C., Sumner, E. J. and Bushby, F. H. (1951). Dynamical methods in synoptic meteorology. Discussion in *Q. J. Roy. Meteor. Soc.*, **77**, 457–73.

Sutherland, I. (1993). See LFR II.

Sutton, O. G. (1954). The development of meteorology as an exact science. *Q. J. Roy. Meteor. Soc.*, **80**, 328–38.

Swarztrauber, P. N. and Kasahara, A. (1985). The vector harmonic analysis of Laplace's tidal equations. *SIAM J. Sci. Stat. Comput.*, **6**, 464–91.

Sweet, R. A. (1977). A cyclic reduction algorithm for solving block tridiagonal systems of arbitrary dimension. *SIAM J. Num. Anal.*, **14**, 706–20.

Swinbank, R., Shutyaev, V. and Lahoz, W. eds. (2003). *Data Assimilation for the Earth System.* NATO Science Series IV, Vol. 26. Kluwer Academic Publishing.

Taba, H. (1988). *The 'Bulletin' Interviews.* World Meteorological Organization, Geneva, WMO No. 708.

Taylor, G. I. (1959). The present position in the theory of turbulent diffusion. *Advances in Geophysics*, **6**, 101–11.

Teweles, S. and Wobus, H. (1954). Verification of prognostic charts. *Bull. Amer. Met. Soc.*, **35**, 455–63.

Thompson, P. D. (1957). Uncertainty of initial state as a factor in the predictability of large scale atmospheric flow patterns. *Tellus*, **9**, 275–95.

Thompson, P. D. (1983). A history of numerical weather prediction in the United States. *Bull. Amer. Met. Soc.*, **64**, 755–69.

Thompson, P. D. (1990). Charney and the revival of NWP. In Lindzen *et al.* (1990), pp. 93–119.

Tijm, A. B. C. (2004). Hirlam pseudo satellite images. Hirlam Newsletter No. 46, SMHI, Norrköping, 59–64.

Uppala, S. M., Kållberg, P. W., Simmons, A. J. *et al.* (2005). The ERA-40 Reanalysis. *Q. J. Roy. Meteor. Soc.*, **131**, 2961–3012.

Wells, D. (1991). *The Penguin Dictionary of Curious and Interesting Geometry.* Penguin Books.

Wiin-Nielsen, A. (1991). The birth of numerical weather prediction. *Tellus*, **43AB**, 36–52.

Wiin-Nielsen, A. (1997). Everybody talks about it . . . . *Mat. Fys. Medd.*, **44:4**, Royal Danish Academy of Sciences and Letters.

Williamson, D. and Temperton, C. (1981). Normal mode initialization for a multilevel gridpoint model. Part II: Nonlinear aspects. *Mon. Weather Rev.*, **109**, 745-57.

Willis, E. P. and Hooke, W. H. (2006). Cleveland Abbe and American meteorology, 1871–1901. *Bull. Amer. Met. Soc.*, **87**, 315–26.

Woods, A. (2005). *Medium-Range Weather Prediction: The European Approach.* Springer.

WPNP: see Richardson, 1922.

Index

Printed in the United States
By Bookmasters